T0199247

# Introduction to Visual Effects

**Introduction to Visual Effects: A Computational Approach** is the first single introduction to computational and mathematical aspects of visual effects incorporating both computer vision and graphics. The book provides the readers with the source code to a library, enabling them to follow the chapters directly and build up a complete visual effects platform. The book covers the basic approaches to camera pose estimation, global illumination, and image-based lighting, and includes chapters on the virtual camera, optimization and computer vision, path tracing, and more.

Key features include:

- Introduction to projective geometry, image-based lighting (IBL), global illumination solved by the Monte Carlo method (*pathtracing*), an explanation of a set of optimization methods, and the techniques used for calibrating one, two, and many cameras, including how to use the RANSAC algorithm in order to make the process robust, and providing code to be implemented using the Gnu Scientific Library.
- C/C++ code using the OpenCV library to be used in the process of tracking points on a movie (an important step for the matchmove process) and in the construction of modeling tools for visual effects.
- A simple model of the bidirectional reflectance distribution function (BRDF) of surfaces and the differential rendering method allowing the reader to generate consistent shadows, supported by a code that can be used in combination with a software like Luminance HDR.

**Bruno Madeira** is a computer engineer from the Military Engineering Institute (IME). He earned his master's and DSc degrees in mathematics with emphasis on computer graphics and vision at IMPA. He has more than 15 years of experience teaching computer graphics at IME and is a member of the Simulators Group at CTEx, Brazil.

**Luiz Velho** is a senior researcher, professor, and a VISGRAF Laboratory leading scientist at the National Institute for Pure and Applied Mathematics (IMPA), Brazil. His academic background includes a BE in industrial design from ESDI-UERJ, an MS in computer graphics from MIT Media Lab, and a PhD in computer science from the University of Toronto. His experience in computational media spans the fields of modeling, rendering, imaging, and animation.

# Introduction to Visual Effects

## A Computational Approach

Bruno Madeira

Luiz Velho

CRC Press is an imprint of the
Taylor & Francis Group, an **informa** business

First edition published 2023
by CRC Press
6000 Broken Sound Parkway NW, Suite 300, Boca Raton, FL 33487-2742

and by CRC Press
4 Park Square, Milton Park, Abingdon, Oxon, OX14 4RN



*CRC Press is an imprint of Taylor & Francis Group, LLC*



ISBN: 978-1-032-07230-2 (hbk)
ISBN: 978-1-032-06124-5 (pbk)
ISBN: 978-1-003-20602-6 (ebk)

DOI: 10.1201/9781003206026

Typeset in Latin Modern font
by KnowledgeWorks Global Ltd.

*Publisher's note:*This book has been prepared from camera-ready copy provided by the authors.

*To our loved ones.*

# Contents

# Preface

This book covers the mathematical and computational aspects for one of the central techniques in visual effects – the camera calibration. Because of its specificity, the subject is usually learned by studying different books and papers about computer vision and graphics that treat individual components separately. Here we propose to offer an integrated view of the involved problems, together with the practical aspects for their solution.

The material presented focuses on the fundamental methods of the area, the implementation problems associated with them, and the relationship between the various components of a computer system for visual effects.

More specifically, the book covers the basic approaches to solve: camera calibration, global illumination, and image-based lighting.

The content is based on the research of Bruno Madeira, who studied the problem of camera pose estimation in his master's dissertation and subsequently extended the results for the creation of visual effects by inserting 3D virtual elements over real 2D videos after his PhD.

In this book, we adopt the "literate programming" paradigm introduced by Donald Knuth. Therefore, besides the theory and algorithms, the book provides alongside the chapters, the source code of a library for implementing the system.

As a basis for the implementation, the visual effects system presented here relies on a 3D graphics library S3D, developed for the book *Design and Implementation of 3D Graphics Systems* [12], written by Luiz Velho. Recently, we have extended this software by adding a path tracing implementation that allows the software to render high-quality scenes.

The implementation of the Pose Estimation uses as a basis the GNU Scientific Library (GSL), an exception occurs in the 2D tracking system and the modeling tools. These softwares have been implemented using routines from the OpenCV Library.

CHAPTER 1

# Introduction

ONE of the central problems that must be solved for the development of visual effects in movies is the calibration problem [1]. In this setting, computer generated three-dimensional objects are integrated seamlessly into the two-dimensional images captured by a camera in a video or film. Therefore, the problem consists in estimating camera parameters used to capture video frames that are needed to combine background images with synthetic objects.

The other two problems that must be solved for generating photo-realistic visual effects combining real images with computer generated 3D graphics are: image-based lighting and global illumination. Besides the information from the live-action camera, it is also necessary to gather data about the light sources present in the scene, since this allows us to produce images of synthetic 3D objects lit in the same condition as the real 3D objects in the scene. Furthermore, the generation of these images must employ algorithms that simulate the physics of light and matter interaction, in other words: global illumination methods.

In the following chapters we will study these three fundamental problems of visual effects: camera calibration; global illumination; and image-based lighting. We will address them from the point of view of Applied Computational Mathematics presenting the mathematical models and computational techniques alongside with the algorithms for implementing their solution.

## 1.1 CAMERA CALIBRATION

In the context of camera calibration, we present a method that combines different computer vision techniques. The solution relies on correspondences from 3D scene points through frame-to-frame associations between 2D image points over the video sequence. Because even short videos are made by hundreds of frames, the correspondence must be done automatically, a process called "tracking." The Kanade-Lucas-Tomasi (KLT) algorithm is used for tracking characteristic points. The method

[1]The term calibration are many times used just for describing the calibration of intrinsic parameters of the camera, but in other books they refers to intrinsic and extrinsic parameters, which is our case.

DOI: 10.1201/9781003206026-1

developed is robust to outliers and assumes that the scene is rigid which makes the camera parameters estimation possible.

Tracking, also known as "match moving," is the process of automatically locating a point or series of points from frame to frame in a sequence, allowing the user to stabilise, track objects or camera movement in the shot [5]. It is the basis for camera calibration and digital compositing [1].

Visual Effects for movies heavily depend on tracking, both in order to combine different 2D images in the same frame as well as to mix synthetic 3D computer graphics elements with real photographic images. In the former case, two-dimensional motion tracking can be incorporated into compositing programs, such as Adobe After Effects and Shake. In the latter case, match moving tools extrapolate three-dimensional information from two-dimensional photography. These are more sophisticated tools that derive camera motion from video sequences and allow using virtual cameras and simulated CG objects in VFX.

Given its importance, most of the book will be devoted to tracking. We will start with a Virtual Camera Model in Chapter 2 and the Optimization Methods to estimate the model from real images in Chapter 3. Subsequently, we will present the basic techniques for computing the model parameters from one and two cameras in Chapters 4 and 5 respectively. Finally, in Chapter 6, we will discuss feature tracking algorithms that allow us to automatically follow points in the 3D scene for a sequence of images. This will make possible to estimate the model parameters from many camera positions in a shot, as presented in Chapter 7.

Complementary with the theory understanding, it is instrumental to review the development of tracking in practice along with the evolution of visual effects in the movie industry, This is the topic of the next section.

## 1.2 HISTORICAL OVERVIEW OF TRACKING

Before the advent of computers in VFX, most effects shots employed optical or electronic methods and required the camera to be fixed in place. Digital tracking opened up many possibilities for modern visual effects, ranging from shot stabilization and addition of matching motion elements to a composite to fully 3D camera animation with complex interaction with virtual objects [24].

The earliest development of tracking for VFX was conducted around 1985 at the NYIT graphics lab by some of the pioneers of Computer Graphics. Later on, ILM (Industrial Light and Magic) developed a 2D tracking software called MM2 that subsequently evolved into the 3D tracking system which was used in films such as Jurassic Park and Star Trek.

Another proprietary system, similarly to ILM's, is the Digital Domain's TRACK system, an integrated software tool that employed computer-vision techniques to extract 2D and 3D information about a scene and the camera. Developed in 1993, the software started as a 2D tracker and evolved into a highly complex 3D camera tracker encompassing sophisticated techniques such as Optical Flow. It has been used over the years by the company in many acclaimed film productions. For this contribution to the film industry, Digital Domain received in 1998 a Technical Achievement Award by the Academy of Motion Picture Arts and Science.

In 1991 Discreet brought tracking to the broader visual effects community with the release of a commercial landmark software called Flame. This system was widely adopted and went through several versions that incorporated news features based on industry demands and the evolution of the state-of-the-art. The first version could do only 1 point tracking. Subsequent versions allowed to track 2-3 or 4 points in the scene with increasingly complex applications, i.e., 1 tracker (stabilization), 2-3 trackers (translation/rotation/scaling) or 4 trackers (corner pinning). Version 5.0, shipped in 1997, could track an arbitrary number of points that made possible automatic triangulation, stabilization, corner tracking, as well as perspective tracking.

Around 1996 two other commercial software systems appeared: Shake and Fusion for Maya. Shake was used for Titanic's opening shot. Maya from Alias/Wavefront was at the time one of the most popular commercial animation systems.

Finally, in 1998 the company REALVIZ was founded aiming to commercialize products based on over ten years of computer vision and robotics research by INRIA. It launched four applications: ImageModeler, ReTimer, Stitcher and MatchMover. These can be considered the next generation of tracking programs.

Nowadays, tracking has become widespread and is incorporated into products such as After Effects (AE) from Adobe. Nonetheless, little is known by the practitioners about the inner algorithms that drive the software. By understanding how motion tracking works, it is possible to improve solutions for VFX.

## 1.3 GLOBAL ILLUMINATION

In computer graphics, to generate the image of a synthetic 3D scene amounts to compute the color information of each point in the image. For this purpose, it is necessary to model three main components of the scene: the camera; the objects and the light sources. The color computation amounts to simulate the physics of light-matter interaction following light rays reaching the camera after being scattered through the scene interacting with the objects and originated at the lights.

Conceptually, we can divide the photo-realistic rendering computation into two parts: local and global illumination. Local illumination takes into account the direct illumination coming from rays emitted by the light sources. These rays interact with the object's materials and are scattered throughout the scene, illuminating other objects until reaching the camera. This process is called Global Illumination, since it operates in the entire scene.

Ray Tracing is the rendering algorithm of choice for photo-realistic visual simulation since it basically implements the optical physics process of light propagation. In Chapter 9 we will study Path Tracing and Monte Carlo methods for Global Illumination.

## 1.4 IMAGE-BASED LIGHTING

For the visualization of 3D scenes only composed of virtual objects, the path tracing algorithm discussed in the previous section is all we need to generate photo-realistic images, assuming that the mathematical models of scene components (camera, objects and lights) have been defined.

However, to create visual effects combining images of a real scene with synthetic 3D objects we need camera calibration and image-based lighting. As we have seen, calibration provides the same parameters for the real and virtual cameras making sure that the images will match geometrically (see [4]). On the other hand, image-based light gives the illumination information from the real scene, such that it can be applied to the virtual lights illuminating the synthetic objects in the scene.

Essentially, image-based lighting captures a 360 degrees representation of the light in the real scene as an omnidirectional image. This representation is called light-map and carries the radiance energy used in the light calculations. For this reason, the light-map must be captured to encode a high-dynamic range radiance information. The seminal contribution in this area was due to Paul Debevec in [2], with a follow up in [3].

The techniques capturing and computing with light-maps will be presented in Chapter 10: Image-Based Lighting.

## 1.5 MATHEMATICAL NOTATIONS

We adopted the following mathematical notations:

1. $a, b \in U$ means that $a \in U$ and $b \in U$.
2. $f : W \subset U \to V$ means that $f : W \to V$ and $W \subset U$.
3. $(X)_n$ is a vector of indexed elements $(X_1, \ldots, X_n)$.
4. $a \approx b$ means that the value $a$ is proximal to $b$.
5. If $M$ is a Matrix $M_{ij}$ is the element of the $i^{th}$ line and $j^{th}$ column.
6. $\nabla f(x)$ is the gradient vector in the $x$ point related to the function $f$.
7. $J_f(x)$ is the Jacobian matrix defined in the point $x$ related to the function $f$.
8. $diag(\lambda_1, \ldots, \lambda_n)$ is a diagonal matrix $M$, such that $M_{ii} = \lambda_i$.
9. $d(x, y)$ is the Euclidean distance between $x$ and $y$.
10. $\mathbb{R}P^n$ is the projective space of dimension $n$.

## 1.6 PROJECTIVE GEOMETRY CONCEPTS

For the understanding of the camera calibration of each frame of a video, it is necessary some knowledge about Projective Geometry. Here we resume the main facts about it that must to be known for a good understanding of this book, that are: Projective Space, Projective Transform, Homographies, Homogeneous Coordinates, Affine Points and Ideal Points.

### 1.6.1 Projective Space

The Projective Space of dimension $n$, denoted by $\mathbb{R}P^n$, is the set of all straight lines in $\mathbb{R}^{n+1}$ that crosses the origin.

As a consequence, a point $p \in \mathbb{R}P^n$ is a class of equivalence $p = (\lambda x_1, \lambda x_2, \ldots, \lambda x_{n+1})$, in which $\lambda \neq 0$. In other words:

$$p = (x_1, \ldots, x_{n+1}) \equiv (\lambda x_1, \lambda x_2, \ldots, \lambda x_{n+1}) = \lambda p.$$

.

For example, the Figure 1.1 presents a point in the $\mathbb{R}P^2$.

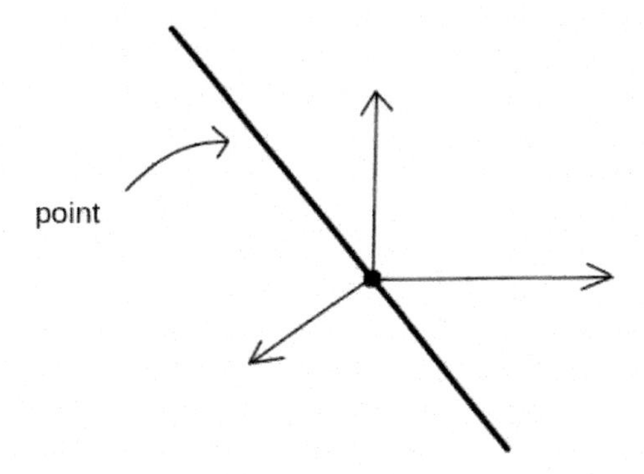


Figure 1.1 A point in $\mathbb{R}P^2$.

This ambiguity of scale means that the projective space is defined by homogeneous coordinates.

The Projective Space $\mathbb{R}P^n$ can be decomposed into two sets: The Affine Space $\mathbb{R}^n$, defined as the set of projective points that can be multiplied to some constant resulting in $x_{n+1} = 1$, and the Space of Ideal Points such that $x_{n+1} = 0$.

In other words:

$$\mathbb{R}P^n = \{(x_1, \ldots, x_{n+1}), x_{n+1} \neq 0\} \cup \{(x_1, \ldots, x_n, 0)\}$$

### 1.6.2 Projective Transforms

A function $T : \mathbb{R}P^n \to \mathbb{R}P^m$ is called Projective Transform if viewed as a function $T : \mathbb{R}^{n+1} \to \mathbb{R}^{m+1}$ is linear.

It means that a Projective Transform $T : \mathbb{R}P^n \to \mathbb{R}P^m$ can be represented by a $(m+1) \times (n+1)$ matrix.

Now we present a theorem that proves that Projective Transformations are invariant to a product to a scalar.

**Theorem 1.1.** *If two Projective Transforms differs uniquely by the multiplication to a scalar value then they represent the same transformation.*

**Proof**

1. By definition $(\lambda T)(x) = \lambda(Tx)$.
2. By the linearity of $T$ , $\lambda(Tx) = T(\lambda x)$.
3. The projective transforms operate over points with homogeneous coordinates. Thus $T(\lambda x) = T(x)$.

Combining these facts we conclude that:

$$T(x) = T(\lambda x) = \lambda(Tx) = (\lambda T)(x).$$

If a Projective Transform is defined over spaces with the same dimension and are invertible, then receive the special name Homography.

### 1.6.3 Projective Geometry on This Book

Many parts of this book use projective geometry, more specifically they use the projective spaces $\mathbb{R}P^2$ and $\mathbb{R}P^3$ for representing points over images or points in the scene.

Coordinates of points projected over images can be both represented as coordinates in $\mathbb{R}^2$ or as homogeneous coordinates of $\mathbb{R}P^2$. The relation between both representations is defined by the transform:

$$(x, y, z)^T \mapsto (\frac{x}{z}, \frac{y}{z})^T,$$

defined when $z \neq 0$.

This is the same as identifying the $\mathbb{R}^2$ with the plane $z = 1$, and considering the intersection between the point in $\mathbb{R}P^2$ with this plane (Figure 1.2).

The same occurs with coordinate points over the 3D scene, they are both specified as points in $\mathbb{R}^3$ or as homogeneous coordinates of affine points in $\mathbb{R}P^3$. It means that the Affine Space of $\mathbb{R}P^n$ can be identified with the $\mathbb{R}^n$ by the relation

$$(x_1, x_2, \ldots, x_{n-1}, x_{n+1}) \mapsto (\frac{x_1}{x_{n+1}}, \frac{x_2}{x_{n+1}}, \ldots, \frac{x_n}{x_{n+1}})^T.$$

In order to avoid confusion, we tried to make clear in which kind of coordinates are been used. For example, in many places we say: "A point $X \in \mathbb{R}^3$ of the scene" or "A point $X \in \mathbb{R}P^3$ of the scene."

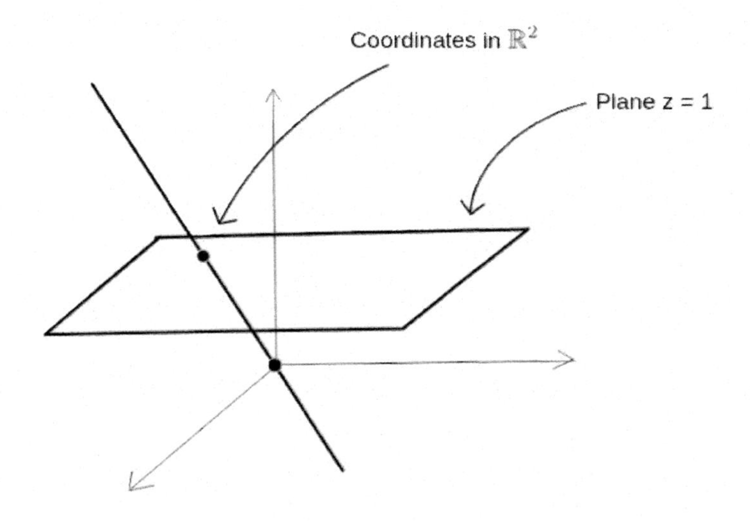


Figure 1.2 Identification of $\mathbb{R}^2$ with the plane $z = 1$.

There are some evaluations of distances between points on this book whose coordinates are specified in the homogeneous form. In other words, $d(x, y)$ defined with $x, y \in \mathbb{R}P^2$. In this case, we assumed that before the evaluation of the distance function, we perform implicitly the coordinate conversion of $x$ and $y$ to $\mathbb{R}^2$, as previously described.

### 1.6.4 Parallelism and Ideal Points

We explained in the previous section the relationship of Affine Points in $\mathbb{R}P^n$ with points in $\mathbb{R}^n$. Now we will explain the meaning of Ideal Points of $\mathbb{R}P^n$ in the $\mathbb{R}^n$ space. We will explain it in the case of $\mathbb{R}P^2$, which can be generalized to an arbitrary dimension. In order to do this, firstly, lets introduce the concept of a line in $\mathbb{R}P^2$.

A line in $\mathbb{R}P^2$ is a plane in $\mathbb{R}^3$ that passes through the origin. Of course, this definition makes a line a set of points in $\mathbb{R}P^2$.

If we analyze the relationship of the Affine points that belongs to a line to coordinates in $\mathbb{R}^2$, we conclude that they are on a straight line. This result is clear because the coordinates in $\mathbb{R}^2$ are the intersection of the plane that define the line in $\mathbb{R}P^2$ with the plane $z = 1$ (Figure 1.3).

Now, lets consider two lines whose affine points intersect on the plane $z = 1$ (Figure 1.4). We conclude that the effect in $\mathbb{R}P^2$ is that the intersection is also a point in $\mathbb{R}P^2$.

An interesting case occurs if the affine coordinates of the lines are parallel. In this case the intersection belongs to the plane $z = 0$, which means that the intersection is an ideal point (Figure 1.5(a)).

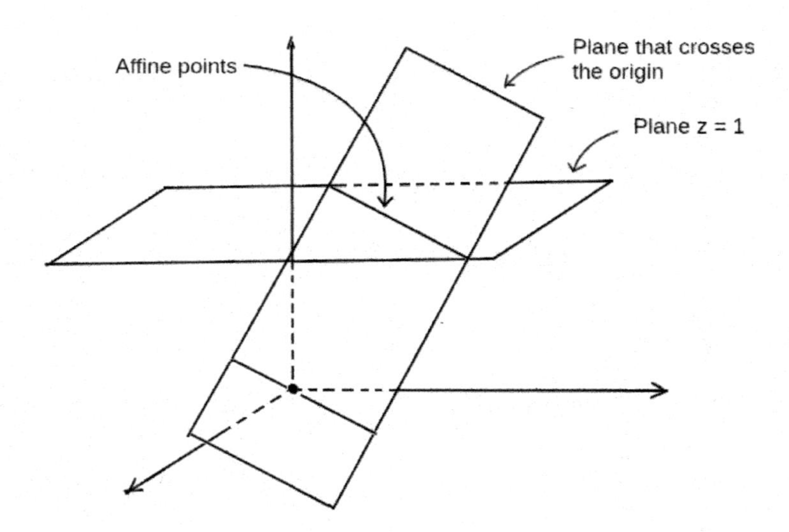


Figure 1.3 The affine points of a line in $\mathbb{R}P^2$ belongs to a line in the plane $z = 1$.

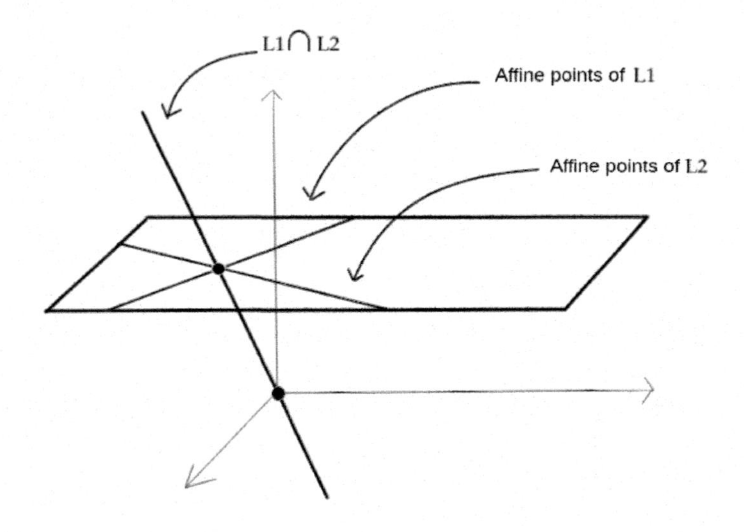


Figure 1.4 The intersection of lines in $\mathbb{R}P^2$ is a point in the plane $z = 1$.

The Figure 1.5(b) shows that if the intersection of two lines in $\mathbb{R}P^2$ is the point $(a, b, 0)^T$ then the direction of parallelism of the affine coordinates is $(a, b)$.

This fact is very important, because it can be generalized to $\mathbb{R}P^3$, which means that if a set of lines in $\mathbb{R}P^3$ intersects on the ideal point $(a, b, c, 0)^T$, we can conclude that lines in Affine Space are parallel lines in the direction $(a, b, c)^T$. This fact will be explored when we talk about the camera model used for rendering in the S3D Library.

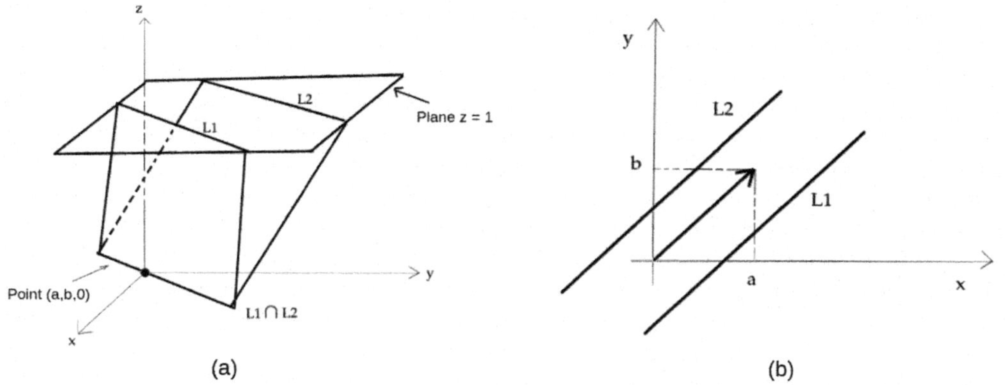

Figure 1.5 (a) Two lines in $\mathbb{R}P^2$ whose affine points are parallel intersecting on an ideal point. (b) The ideal point intersection defines the direction of parallelism.

## 1.7 ABOUT THE CODE

The book describes two code libraries available in the site https://github.com/visgraf/vfx: one implements a MatchMove software and the other implements a Path Tracing software.

The first one is mostly constructed using mathematical routines from the GNU Scientific Library, excluding the implementation of the 2D tracker and the Modeling Tool, which have been implemented using routines from the OpenCV Library.

The second library is an extension of the S3D Library, presented in the book "Design and Implementation of 3D Graphics Systems" [12].

The reader must consider that the code in this book have only the purpose to be a didactic tool, since they are all implemented in a naive way. Although the MatchMove software can be used to estimate the pose of the camera of a small video, the Path Tracing cannot generate a low-noise image in a reasonable time. As a consequence, if the reader wants to generate high-quality images, we encouraged him to generate the visual effect over chosen frames in the video, but not for a whole sequence of frames.

The reader that is only interested in testing the MatchMove software can use it to write a plugin for a high-quality render software such as Maya or Blender.

As an example, we developed a plugin for Maya and used it to generate the video illustrated in the frames presented in Figure 1.6.

Figure 1.6 Frames of a video generated by a plugin written for Maya in which a virtual sphere and a virtual cube are added to the scene.

CHAPTER 2

# Virtual Camera

A VIRTUAL CAMERA is a mathematical object that describes the operation of an optical camera; that is, it establishes the correspondence between elements of the three-dimensional world and its projections on an image. In the context of Visual Effects, we need camera models to solve two types of problems:

1. Image synthesis problems.
2. Calibration problems.

In this chapter, we will address camera models appropriate for the resolution of these problems. Initially, we will present a basic camera model, which will be defined without making use of projective geometry. This will serve as a basis for the definition of two models created with projective geometry, one being used for solving projects problems of image synthesis, and another used to solve calibration problems. Parameterizations for these models will be presented. Similar to other texts, we will group the parameters into two categories: extrinsic parameters and intrinsic parameters.

The extrinsic parameters describe the positioning and orientation of the camera.

The intrinsic parameters, on the other hand, describe the effect of the camera on the luminous rays and the action of the camera's sensors in the formation of the image. The properties of camera controlled by intrinsic parameters include: focal length, resolution of the image, the dimensions of the pixels, the radial distortion caused by the lens, ... etc.

The mapping of camera models used in calibration on the models used in image synthesis will be done implicitly, by adopting the same nomenclature in the parameterization of both. For example, the letter $d$ will be used to specify the focal length both in models used in image synthesis and in models used in calibration.

At the end of the chapter, we will explain how to map the extrinsic and intrinsic parameters to S3D library functions.

We highlight that this section describes the relationship between the camera model used in Computer Vision and in Computer Graphics in general. It means that, although the Visual Effects uses the Path Tracing algorithm, that do not need a camera model able to perform operations such as clipping; we decided to consider a

DOI: 10.1201/9781003206026-2

model based on the S3D Library, which can be used both in a Path Tracing software, or in a modeler software able to render in real time. For more details consult [12].

## 2.1 BASIC MODEL

A basic camera model will now be presented, which will be later specialized in solving problems of image synthesis and calibration. A hypothesis adopted throughout the text is that the effect on light rays produced by a camera that has lenses, can be approximated by the effect produced by a pinhole camera [9], which is the type of camera considered in the models that will be presented. A more general treatment, which takes into account the radial distortion caused by the lens, can be found in [4].

A pinhole camera performs a perspective projection of the points in a scene over a plane. As the optical center of the camera is located between the projection plane and the projected objects, an inversion of the captured image occurs. Since this does not generate much problems from a mathematical point of view, it is common to describe the effect of a pinhole by a perspective projection in which the projection plane is between the projection center and the projected objects, obtaining an equivalent result without the inversion, as illustrated in Figure 2.1.

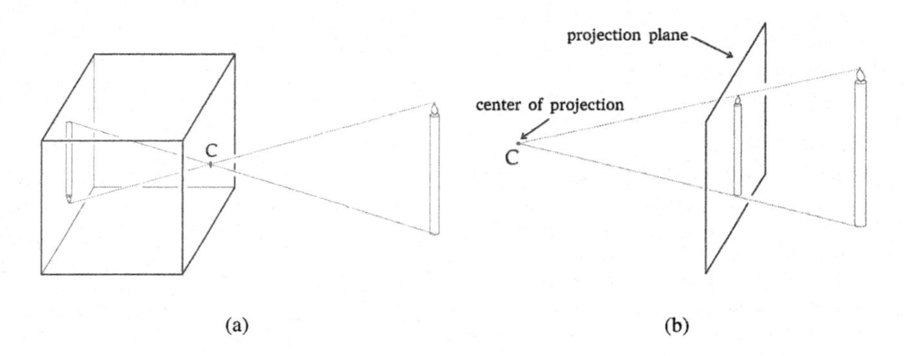


Figure 2.1 (a) Pinhole camera; (b) camera model.

Next, three transformations will be defined, called $T_1$, $T_2$ and $T_3$, which will be combined to form the basic camera model.

### 2.1.1 Camera in the Origin

For a perspective projection whose projection center is positioned at $(0,0,0)^T$, and whose projection plane is perpendicular to the z-axis, the transformation associated is $T_1 : S \subset \mathbb{R}^3 \to \mathbb{R}^2$, defined by

$$T_1(x, y, z)^T = \left(d\frac{x}{z}, d\frac{y}{z}\right), \tag{2.1}$$

in which $S$ is the set formed by the points of $\mathbb{R}^3$ that do not have $z = 0$, and $d$ corresponds to the distance between the center and the projection plane. This distance is called focal length.

### 2.1.2 Camera in Generic Position

The transformation corresponding to an arbitrarily positioned camera is given by the composition $T_1 \circ T_2 : T_2^{-1}(S) \rightarrow \mathbb{R}^2$, such that $T_2 : \mathbb{R}^3 \rightarrow \mathbb{R}^3$ is a rigid movement defined by

$$T_2(x) = R(x - c), \tag{2.2}$$

in which $c$ is the position of the projection center and $R$ is a rotation matrix, which determines the orientation of the camera.

The rotation matrix $R$ and the vector $c$ can be parameterized by 6 real numbers, which correspond to the extrinsic parameters of the camera.

### 2.1.3 Digital Camera

In the case of digital cameras, the image is projected on a matrix sensor, which perform a sampling of the same. This sampling defines a new coordinate system for the projected image. Changing coordinates of the image is defined by an affine transformation in the plane $T_3 : \mathbb{R}^2 \rightarrow \mathbb{R}^2$, of the form,

$$T_3(x) = diag(m_x, m_y) + (x_0, y_0)^T, \tag{2.3}$$

such that $m_x$ and $m_y$ correspond to the number of sensors per unit length in the direction $x$ and $y$, respectively, and the pair $(x_0, y_0)^T$ corresponds to the principal point, which defines the pixel scale coordinates of the orthogonal projection of the projection center about the projection plane.

### 2.1.4 Intrinsic Parameters

We will now analyze the composition $T_3 \circ T_1 : S \rightarrow \mathbb{R}^2$. This transformation is defined by

$$T_3 \circ T_1\{(X, Y, Z)^T\} = \left(dm_x \frac{x}{z} + x_0, dm_y \frac{y}{z} + y_0\right)^T. \tag{2.4}$$

It is immediate to verify, by the expression above, that digital cameras with different focal points distances can produce the same result, simply by choosing an appropriate spatial resolution. It happens because these values appear combined in the $dm_x$ and $dm_y$ products. The values $x_0$, $y_0$, $dm_x$ and $dm_y$ define the intrinsic parameters of the basic camera model.

### 2.1.5 Dimension of the Space of Virtual Cameras

The transformations $T_3 \circ T_1 \circ T_2 : T_2^{-1}(S) \rightarrow \mathbb{R}^2$ defines the space of virtual cameras that have 10 degrees of freedom, 3 degrees of freedom associated the rotation $R$, 3 degrees of freedom associated with the position of the projection center $c$, and the others 4 degrees of freedoms defined by intrinsic parameters.

We highlight that when we are considering a model with 10 degrees of freedom, we are disregarding that the dimensions of the plane of projection are intrinsic parameters. From the point of view of calibration, this does not create any problem, because the limitations of the screen are being physically applied by the camera.

On the other hand, from the point of view of image synthesis, these dimensions are important.

## 2.2 CAMERA FOR IMAGE SYNTHESIS

The problem of image synthesis can be defined as that of creating images from three-dimensional scene descriptions. This section and the next deal with a camera model suitable for image synthesis.

### 2.2.1 Terminologies

The main terms used in the specification of cameras will now be presented in computer graphics. Figure 2.2 illustrates each one.

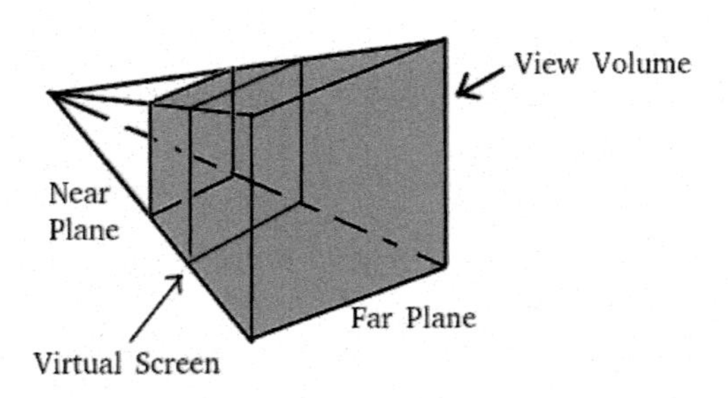


Figure 2.2 View Pyramid.

**Virtual screen** is the rectangle of the projection plane that contains the projected image. That limitation defined on the projection plane physically corresponds to the limitations in the dimensions of the screen where the light rays are projected.

**View pyramid** is the pyramid defined by the projection center and the virtual screen. Near plane is a plane positioned in front of the projection center. Points only that are in front of the near plane are projected on the image. Far plane is a plane positioned in front of the near plane. Points only that are behind the far plane are projected on the image.

**View volume** is the pyramid trunk defined by the portion of the view pyramid bounded by the near plane and the far plane.

### 2.2.2 Clipping and Visibility

The basic camera model defined by the transformation $T_3 \circ T_1 \circ T_2 : T_2^{-1}(S) \to \mathbb{R}^2$ is able to describe the position in the image of all the points in the scene that are projected. On the other hand, he defines projections for points in the scene that would not be projected by the corresponding pinhole camera. More precisely, a point in the scene $\mathbb{X} \in \mathbb{R}^3$ to be projected by a camera must satisfy the following properties:

1. $X$ must be in front of the camera.
2. The projection of $X$ must be contained in the camera's screen.
3. $X$ must not be occluded from another point in the scene.

The View Pyramid is the geometrical place of the points that satisfy the properties 1 and 2. The determination of the points of the scene that belong to the vision pyramid is called clipping in relation to the view pyramid. The problem of determining the points that satisfy property 3 is known as the visibility problem.

In Computer Graphics, it is required, in addition to these three properties, that $X$ belongs to the region of space bounded by the near and far planes, replacing the operation of clipping in relation to the viewing pyramid by clipping in relation to the viewing volume.

The objective of the constraint given by the near plan is to avoid numerical problems when dividing by very small numbers. This type of error can occur, for example, if we apply the transformation $T_1$, defined by Equation 2.1, to a point very close to the projection center.

The objective of the constraint given by the far plane is to limit the depth of the region of the scene that will be projected, allowing you to use the Z-buffer algorithm to solve visibility problems.

An analysis of algorithms that solve problems of visibility and clipping are outside the scope of this book. This subject is discussed in detail in [12].

## 2.3 TRANSFORMATION OF VISUALIZATION

Usually, in the image synthesis process, there are used a camera model formed by a sequence of projective transformations in $\mathbb{R}P^3$ interspersed with algorithms that solve the problems of clipping and visibility. We will deal with a particular sequence adopted by the book [12] (Figure 2.3), where the S3D Library is defined. More precisely, we will make an adaptation of the model defined in [12] to the notation established in Section 2.1.

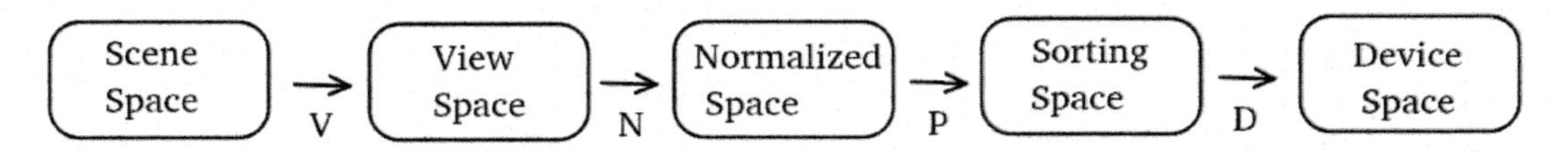


Figure 2.3 Transforms that compound the camera model used in image synthesis.

### 2.3.1 Positioning the Camera

Transformation $V : \mathbb{R}P^3 \to \mathbb{R}P^3$ changes the coordinate system of the scene for the camera's coordinate system, that is, it is a projective version for the rigid motion defined by $T_2$ in Section 2.1.2. Its matrix representation is

$$V = \begin{pmatrix} R & -Rc \\ 0^T & 1 \end{pmatrix} \tag{2.5}$$

### 2.3.2 Transformation of Normalization

The transformation $N : \mathbb{R}P^3 \to \mathbb{R}P^3$ maps the coordinate system of the camera into a standardized coordinate system, where the clipping problem is simplified. Its matrix representation is

$$N = \begin{pmatrix} \frac{d}{fs_x} & 0 & \frac{x_0 - m_x s_x}{f m_x s_x} & 0 \\ 0 & \frac{d}{fs_y} & \frac{y_0 - m_y s_y}{f m_y s_y} & 0 \\ 0 & 0 & \frac{1}{f} & 0 \\ 0 & 0 & 0 & 1 \end{pmatrix} \tag{2.6}$$

where we have, respectively, that $n$ and $f$ are the distances from the near and far planes to the projection center, and $2s_x$ and $2s_y$ are the horizontal and vertical dimensions of the virtual screen.

The clipping problem in relation to the vision pyramid is simplified, because in the normalized coordinate system, the view pyramid is mapped to the pyramid defined as

$$\{(x, y, z)^T \in \mathbb{R}^3 : -z < x < z, -z < y < z, 0 < z\}$$

### 2.3.3 Perspective Projection

The transformation $P : \mathbb{R}P^3 \to \mathbb{R}P^3$ maps the coordinate system normalized into the sorting coordinate system. Its matrix representation is defined by

$$P = \begin{pmatrix} 1 & 0 & 0 & 0 \\ 0 & 1 & 0 & 0 \\ 0 & 0 & \frac{f}{f-n} & \frac{-n}{f-n} \\ 0 & 0 & 0 & 1 \end{pmatrix}. \tag{2.7}$$

When describing the scene in the sorting coordinate system, we obtain two interesting properties, which are:

1. In this referential, when applying a transformation $\Pi : \mathbb{R}P^3 \to \mathbb{R}^2$ , defined by $(a_x, a_y, a_z, 1)^T \mapsto (a_x, a_y)$, we obtain the perspective projection made by the corresponding virtual camera.

2. In this referential, a point $A = (a_x, a_y, a_z, 1)^T$ T exerts an occlusion on a point $B = (b_x, b_y, b_z, 1)^T$, if and only if, $\Pi(A) = \Pi(B)$ and $a_z < b_z$.

These two properties show that both the perspective calculation and the solution to the visibility problem can be carried out in a trivial way in the sorting coordinate system.

### 2.3.4 Device Coordinates

Transformation $D : \mathbb{R}P^3 \to \mathbb{R}P^3$ maps the coordinate system of sorting into the device's coordinate system. This coordinate system has some interesting properties:

1. The two properties of the ordering reference are still valid.
2. The coordinates of the $x$ and $y$ axes given in pixel scale.
3. The volume of vision, in the direction of the z-axis, corresponds exactly to the representation of the Z-buffer.

The matrix representation of transformation D is given by:

$$D = \begin{pmatrix} s_x m_x & 0 & 0 & s_x m_x \\ 0 & s_y m_y & 0 & s_y m_y \\ 0 & 0 & Z_{max} & 0 \\ 0 & 0 & 0 & 1 \end{pmatrix} \quad (2.8)$$

in which $[0, Z_{max}]$ is the representation interval of the Z-buffer.

## 2.4 COMPARISON WITH THE BASIC MODEL

### 2.4.1 Intrinsic Parameters

The matrix associated with the DPN composition is given by

$$DPN = \begin{pmatrix} dm_x & 0 & x_0 & 0 \\ 0 & dm_y & y_0 & 0 \\ 0 & 0 & \frac{fZmax}{f-n} & \frac{-nfZmax}{f-n} \\ 0 & 0 & 1 & 0 \end{pmatrix} \quad (2.9)$$

The restriction of the DPN image to the xy plane is equal to the effect of the transformation $T_3 \circ T_1$ defined in by Equation 2.4. More precisely, $\Pi \circ DPN = T_3 \circ T_1$, in which $\Pi : \mathbb{R}P^3 \to \mathbb{R}^2$ is defined as in Section 2.3.3. The effect of DPN in the direction of the z-axis is an affine adjustment of the View Volume over the interval $[0, Z_{max}]$.

This matrix does not appear explicitly in a real-time graphic system, since you cannot compose DP with N, because performing the clipping operation after the application of N is necessary. But the composition is interesting, because it makes evident that the effect of the intrinsic parameters $x_0, y_0, dm_x$ and $dm_y$ are the same as in the basic model. In addition, this matrix displays two extra intrinsic parameters $n$ and $f$, that do not correspond to real-world camera parameters, and whose effect on the image generated is the elimination of projected surfaces.

The $Z_{max}$ value is not determined by the state of the camera, but by the device used by the Z-buffer algorithm, therefore, it is not a camera parameter, and does not cause any influence on the image generated by the model.

Another important observation is that $DPN$ is free in $s_x$ and $s_y$ , which is a fact expected, since these values do not appear in $T_3 \circ T_1$ . However $s_x$ and $s_y$ are relevant values because they define the dimensions of the image, therefore they are intrinsic parameters of the camera.

We also must observe what this transform makes to the origin of the camera, represented by the coordinates $(0, 0, 0, 1)^T$ in the domain of the $DPN$ transform.

We can easily observe that $DPN(0,0,0,1)^T = (0,0,1,0)^T$. The meaning of this is that the rays captured by the camera, converging to the origin (optical center), after applying $DPN$ becomes parallel to the direction of the z-axis.

### 2.4.2 Dimension

It is concluded that the camera model used in image synthesis has 14 degrees of freedom. In addition to the 10 degrees of freedom of the basic model, there are four other intrinsic parameters, with two corresponding to the dimensions of the virtual screen and the two others corresponding to the distances from the projection center to the near and far planes.

### 2.4.3 Advantages over the Basic Model

The reasons that make the use of projective transformations in $\mathbb{R}P^3$, in the construction of graphic systems, instead of the formulation made in the basic model are the following:

1. Projective transformations in $\mathbb{R}P^3$ allow to represent both rigid movements in space as projection operations.

2. The visibility problem is simplified by choosing a coordination system appropriate in $\mathbb{R}P^3$, such as in the case of sorting coordinate systems and in the device coordinate system. It happens because this problem can be is easily solved by using the Z-buffer algorithm in the case of a real-time graphic system [12].

## 2.5 CAMERA FOR PATH TRACING

The Path Tracing is an algorithm focused on the image [12]. As consequence, we may use the inverse of Projective Transforms used for real-time graphics system in the inverse order (Figure 2.4).

$$\text{Image} \longrightarrow D^{-1} \longrightarrow P^{-1} \longrightarrow N^{-1} \longrightarrow V^{-1} \longrightarrow \text{3D Scene}$$

Figure 2.4 Sequence of projective transforms used by the path tracing.

Another important fact is that path tracing does not use the clipping algorithm and does not use the Z-buffer for solving the visibility problem. As a consequence, all the matrix can be composed in a single projective transform that transform coordinates in the device space into coordinates in the scene space.

## 2.6 VISIBILITY AND RAY CASTING

As mentioned in the previous section, Path Tracing is an algorithm based in the image. It uses Ray Casting as the central mechanism. In order to understand the role of visibility operation for Ray Casting it helps to compare the differences of viewing transformations in the two main visualization approaches: i.e., algorithms based in the scene and algorithms based in the image.

In the scene-based approach, points $x \in \mathbb{R}^3$ belonging to objects of the 3D scene are transformed by applying the sequence of viewing transformations directly, e.g.,: $D \circ P \circ N \circ V(x)$ and the visibility is solved in the sorting space.

In the image-based approach, rays $r(t)$ originating from the camera are transformed by applying the sequence of viewing transformations in inverse order, as described in Figure 2.4. In this case, the visibility is solved by comparing the ray parameter $t$ for points $p_i = r(t_i)$ along the ray trajectory. These points, correspond to the intersection of the ray with the objects $\mathcal{O}_i$ in the scene, an operation that is computed in the local space of the object. Since the parameter $t$ ranges in the interval $[0, \infty]$ starting at the camera, the visible points at each pixel are the ones with smallest $t$ values.

The S3D library employs a unified model of the viewing transformations for both scene and image-based approaches. The differences, as we discussed above, are that in the scene-based approach direct transformations are applied for points of the scene and in the image-based approach inverse transformations applied to camera rays.

In this respect, rays, $r(t) = o + td$, are defined in the scene space by their origin, $o$ and direction, $d$. They are transformed to image/device space to form the sampling pixel grid and subsequently to the local space of each object in the scene for visibility computation. Since the parameter $t$ is invariant by these transformations, the visibility remains intrinsic to each ray.

Of course, we must take into account is that $o$, in the device space, is the ideal point correspondent to the direction defined by the z-axis. As a consequence, the rays $r(t) = o + td$ must be traced parallel to this direction when we are using the S3D Library.

## 2.7 CAMERAS FOR CALIBRATION

The calibration problem consists in determining the intrinsic parameters and extrinsic parameters of a set of cameras.

This generic problem can be specialized in different modalities. In our case we are interested in the following formulation in particular:

Given a set of $n$ images, determine the extrinsic and intrinsic parameters of the $n$ cameras that captured these images. In this case, we will say that the $n$ cameras are consistent with the $n$ images, and that the $n$ cameras provide an explanation for the $n$ images. In practice, the calibration problem is formulated from the point of view of optimization, since measurement errors in images generally makes the problem of defining a consistent set of cameras impossible. Thus, the calibration problem starts to be reformulated as: Given a set of $n$ images, determine the extrinsic and intrinsic parameters of the $n$ cameras that best explain the $n$ images. The mathematical formalization of this optimization problem, and an algorithm that solves it, is presented in Chapter 7.

### 2.7.1 Projective Model

The camera model used in calibration can be obtained by rewriting the transformations $T_1$, $T_2$ and $T_3$ defined in Section 2.1 as projective transformations $T_1 : \mathbb{R}P^3 \to \mathbb{R}P^2$, $T_2 : \mathbb{R}P^3 \to \mathbb{R}P^3$ and $T_3 : \mathbb{R}P^2 \to \mathbb{R}P^2$, obtaining the following matrix representations:

$$T_1 = \begin{pmatrix} d & 0 & 0 & 0 \\ 0 & d & 0 & 0 \\ 0 & 0 & 1 & 0 \end{pmatrix}, T_2 = \begin{pmatrix} R & -Rc \\ 0^T & 1 \end{pmatrix} T_3 = \begin{pmatrix} m_x & 0 & x_0 \\ 0 & m_y & y_0 \\ 0 & 0 & 1 \end{pmatrix}$$

### 2.7.2 Projective Notation for Cameras

It is immediate to verify that the projective transformations $T_3 \circ T_1 \circ T_2 : \mathbb{R}P^3 \to \mathbb{R}P^2$ can be represented by the product of a $3 \times 3$ matrix by a $3 \times 4$ matrix, as shown below:

$$T_3 \circ T_1 \circ T_2 = \begin{pmatrix} dm_x & 0 & x_0 \\ 0 & dm_y & y_0 \\ 0 & 0 & 1 \end{pmatrix} \begin{pmatrix} R & -Rc \end{pmatrix}. \tag{2.10}$$

In this case, it is common to use the compact notation $K[R| - Rc]$ to express this product. In this notation, $K$ corresponds to the $3 \times 3$ matrix that specifies the intrinsic parameters of the camera, and $[R| - Rc]$ corresponds to the $4 \times 3$ matrix that specifies the extrinsic parameters . It is also common to use the $K[R|t]$ notation, whose only difference for previous notation is that the position of the projection center is not explicit, because the product $-Rc$ is replaced by a vector $t \in \mathbb{R}^3$, which represents the translation of camera.

### 2.7.3 Generic Projective Camera

We have that the transformations $T_3 \circ T_1 \circ T_2 : \mathbb{R}P^3 \to \mathbb{R}P^2$ defines a set of $4 \times 3$ matrices that have 10 degrees of freedom. Considering that the set formed for all the projective transformations defined in $\mathbb{R}P^3 \to \mathbb{R}P^2$ has 11 degrees of freedom, it is concluded that there are certainly projective transformations of this set that do not correspond to any camera.

It will be shown, in the Chapter 4, that this extra degree of freedom can be obtained considering a model for cameras defined by projective transformations of the form

$$\begin{pmatrix} f_1 & s & x_0 \\ 0 & f_2 & y_0 \\ 0 & 0 & 1 \end{pmatrix} \begin{pmatrix} R & -Rc \end{pmatrix}.$$

This model features a generic projective camera [13], which has five intrinsic parameters: $f_1$ , $f_2$ , $s$, $x_0$ and $y_0$. The extra degree of freedom allows the angle $\theta$, defined by the $x$ and $y$ axes, which specify the image coordinate system, can be modified. Physically this can be interpreted as a shear in the matrix of sensors from a digital camera. The parameters $f_1$, $f_2$ and $s$ are related to the parameters of the 10-degree of freedom model [9]:

$$f_1 = dm_x, \tag{2.11}$$

$$f_2 = \frac{dm_y}{sin\theta}, \tag{2.12}$$

$$s = -f_1 cotg\theta. \tag{2.13}$$

The pair $(x_0, y_0)^T$ has the same interpretation as the 10 degrees of freedom model, specifying the coordinates, in pixel scale, of the principal point.

## 2.8 MAPPING A CALIBRATED CAMERA INTO THE S3D LIBRARY

We have shown in the previous sections how intrinsic and extrinsic parameters are inserted in the projective transformations that make up models of cameras used in camera calibration and in image synthesis. We will now present how these parameters can be used to specify a camera in the S3D library. More precisely, we will show the S3D library function calls necessary to define the parameters of a $K[R|t]$ camera, possibly estimated by a calibration process.

We highlight that many others render systems can have their intrinsic and extrinsic parameters of cameras adjusted in the same way as S3D.

### 2.8.1 Specification of Extrinsic Parameters

The S3D Library has a structure Camera that contains a field $V$ whose type is Matrix4. This field is the $4 \times 4$ matrix that represents the extrinsic parameter of the camera. More precisely, we must have:

$$V = \begin{pmatrix} R & t \\ 0^T & 1 \end{pmatrix},$$

such that $0^T = (0, 0, 0)$.

### 2.8.2 Specification of Intrinsic Parameters

Specifying intrinsic parameters is less immediate. We observed especially that the cameras defined by S3D do not present shear in the matrix sensors, that is, if we want to specify the intrinsic parameters of a camera $K[R|t]$ it is necessary that the matrix $K$ have form

$$K = \begin{pmatrix} f_1 & 0 & u_0 \\ 0 & f_2 & v_0 \\ 0 & 0 & 1 \end{pmatrix}. \tag{2.14}$$

In this case, you can use the function *frustum*, whose prototype is defined by

```
void frustum(View *v, Real l, Real b, Real r, Real t,
             Real near, Real far)
```

The arguments of the frustum function define in the camera reference the coordinates in $\mathbb{R}^3$ of the lower left and upper right vertices of the virtual screen, as being $(l, b, near)^T$ and $(r, t, near)^T$ respectively. The $near$ and $far$ parameters defines the distance from the projection center to the near and far planes. In addition, the $near$ plane coincides with the projection plane, that is, the focal distance is the $near$ value.

We need to determine the arguments that must be passed to frustum's so that the viewing volume is compatible with the intrinsic parameters of the $K$ matrix.

Observing the Equations 2.11 and 2.12, we have that the number of sensors per unit horizontal and vertical length, measured on the virtual screen, are defined respectively by

$$m_x = \frac{f_1}{near}, \tag{2.15}$$

$$m_y = \frac{f_2}{near}. \tag{2.16}$$

The coordinates of the principal point, measured in the image coordinate system, are $(u_0, v_0)^T$. As the coordinates of the principal point on the virtual screen are $(0, 0, near)^T$, it is concluded that the Frustum function must be called by passing the following arguments:

$$l = -\frac{u_0}{m_x}, r = \frac{w - u_0}{m_x}, b = -\frac{v_0}{m_y}, t = \frac{h - v_0}{m_y},$$

where $w$ and $h$ correspond respectively to the horizontal and vertical resolution of the image captured by the camera.

## 2.9 API

### 2.9.1 MatchMove Software Functions

Here we present the API of functions defined in the MatchMove software to manipulate cameras.

```
void calib_pmatrix_make( gsl_matrix *p, gsl_matrix *k,
                         gsl_matrix *r, gsl_vector *t );
```

This function returns in $p$ the camera $k[R|t]$.

```
void calib_apply_P( gsl_matrix *p, gsl_vector *x,
                    gsl_vector *x_proj );
```

This function applies the camera matrix $p$ to $x$ and returns the projection in $x_{proj}$.

```
gsl_matrix *calib_kmatrix_alloc( double f1, double f2,
                                 double s, double u,
                                 double v );
```

This function returns an intrinsic parameter of a camera with the form

$$\begin{pmatrix} f1 & s & u \\ 0 & f2 & v \\ 0 & 0 & 1 \end{pmatrix}$$

### 2.9.2 Render Software Functions

Here we present the API of functions defined in the render software to manipulate cameras.

```
CamData *cam_data_alloc( int ncams, int w, int h,
                          Real near, Real far );
```

This function creates a structure *CamData* that is able to represent the extrinsic parameter of *ncams* cameras. It also defines the image resolution of the camera and the near and far planes of the camera.

```
void cam_data_free( CamData *cd );
```

This function destroys a structure *CamData*.

```
Matrix3  kmatrix_read( FILE *f );
```

This function returns an intrinsic parameter matrix specified by reading the file $f$, written by the program described in the Section 4.11.

```
Matrix4 rtmatrix_read( FILE *f );
```

This function returns an extrinsic parameter matrix specified by reading the file $f$, written by the programs described in the Sections 7.16 and 7.17.

```
void mmove_read( CamData  *cd, char *kfname, char *rtfname );
```

This function returns in the parameter *cd* a *CamData* structure specified by the file named $kfname$, which defines the intrinsic parameter, and by the file named $rtfname$, which defines the extrinsic parameters.

```
void mmove_view( View *view, CamData *cd, int frame );
```

This function defines the *view* as being the one specified by the frame $frame$ of the structure *CamData cd*.

## 2.10 CODE

### 2.10.1 Code in the MatchMove Software

```
// calib/pmatrix.c

void calib_pmatrix_make( gsl_matrix *p, gsl_matrix *k, gsl_matrix *r,
   gsl_vector *t )
{
 int i;
 gsl_vector *v;
 gsl_matrix *paux;

 v = gsl_vector_alloc(3);
 paux = gsl_matrix_alloc(3,4);

 for( i=0; i<3; i++ ){
   gsl_matrix_get_col( v, r, i );
   gsl_matrix_set_col( paux, i, v );
 }

 gsl_matrix_set_col( paux, 3, t );
 gsl_linalg_matmult( k, paux, p );
 gsl_vector_free(v);
 gsl_matrix_free(paux);
}

// calib/applyP.c

void calib_apply_P( gsl_matrix *p, gsl_vector *x, gsl_vector *x_proj )
{
 gsl_matrix *x_homog;
 gsl_matrix *proj_homog;

 x_homog = gsl_matrix_alloc( 4, 1 );
 proj_homog = gsl_matrix_alloc( 3, 1 );

 gsl_matrix_set( x_homog, 0, 0, gsl_vector_get(x,0) );
 gsl_matrix_set( x_homog, 1, 0, gsl_vector_get(x,1) );
 gsl_matrix_set( x_homog, 2, 0, gsl_vector_get(x,2) );
 gsl_matrix_set( x_homog, 3, 0, 1. );
 gsl_linalg_matmult( p, x_homog, proj_homog );
 gsl_vector_set( x_proj, 0, gsl_matrix_get( proj_homog, 0, 0 )/
                            gsl_matrix_get( proj_homog, 2, 0 ) );
 gsl_vector_set( x_proj, 1, gsl_matrix_get( proj_homog, 1, 0 )/
                            gsl_matrix_get( proj_homog, 2, 0 ) );

 gsl_matrix_free( x_homog );
 gsl_matrix_free( proj_homog );
}

// calib/kmatrix.c

gsl_matrix *calib_kmatrix_alloc( double f1, double f2, double s, double u,
   double v )
{
 gsl_matrix *r;

 r = gsl_matrix_alloc( 3, 3 );
 gsl_matrix_set_zero(r);
 gsl_matrix_set( r, 0, 0, f1 );
 gsl_matrix_set( r, 0, 1, s );
 gsl_matrix_set( r, 0, 2, u );
```

```
 gsl_matrix_set( r, 1, 1, f2 );
 gsl_matrix_set( r, 1, 2, v );
 gsl_matrix_set( r, 2, 2, 1.);

 return r;
}
```

## 2.10.2 Code in the Render Software

```
// lmmove/mmove.h

#ifndef MMOVE_H
#define MMOVE_H

#include "scene.h"
#include "view.h"
#include "image.h"

typedef struct Matrix3 {
  Vector3 r1, r2, r3;
} Matrix3;


typedef struct CamData{
  Real w, h;
  Real near, far;
  int nframes;
  Matrix3 k;
  Matrix4 *rt;
} CamData;


Matrix3  kmatrix_read( FILE *f );
Matrix4 rtmatrix_read( FILE *f );
void mmove_read( CamData  *cd, char *kfname, char *rtfname );
void mmove_view( View *view, CamData *cd, int frame );
CamData *cam_data_alloc( int ncams, int w, int h, Real near, Real far );
void cam_data_free( CamData *cd );

void mmimg_putc( Image *img, int u, int v, Color c  );
Color mmimg_getc( Image *img, int u, int v );


#endif

// lmmove/cdata.c

CamData *cam_data_alloc( int ncams, int w, int h, Real near, Real far )
{
 CamData *cd;

 cd = NEWSTRUCT( CamData );
 cd->w = w;
 cd->h = h;
 cd->near = near;
 cd->far = far;
 cd->rt = NEWTARRAY( ncams, Matrix4 );
 return cd;
}

void cam_data_free( CamData *cd )
```

```
{
 free(cd->rt);
 free(cd);
}

// lmmove/mmove.c

#include "mmove.h"

Matrix3  kmatrix_read( FILE *f )
{
 Matrix3 k;

 fscanf( f, "%lf", &k.r1.x );
 fscanf( f, "%lf", &k.r1.y );
 fscanf( f, "%lf", &k.r1.z );
 fscanf( f, "%lf", &k.r2.x );
 fscanf( f, "%lf", &k.r2.y );
 fscanf( f, "%lf", &k.r2.z );
 fscanf( f, "%lf", &k.r3.x );
 fscanf( f, "%lf", &k.r3.y );
 fscanf( f, "%lf", &k.r3.z );

 return k;
}


Matrix4 rtmatrix_read( FILE *f )
{
 Matrix4 m = m4_ident();

 fscanf( f, "%lf", &m.r1.x );
 fscanf( f, "%lf", &m.r1.y );
 fscanf( f, "%lf", &m.r1.z );
 fscanf( f, "%lf", &m.r2.x );
 fscanf( f, "%lf", &m.r2.y );
 fscanf( f, "%lf", &m.r2.z );
 fscanf( f, "%lf", &m.r3.x );
 fscanf( f, "%lf", &m.r3.y );
 fscanf( f, "%lf", &m.r3.z );
 fscanf( f, "\n" );
 fscanf( f, "%lf", &m.r1.w );
 fscanf( f, "%lf", &m.r2.w );
 fscanf( f, "%lf", &m.r3.w );
 fscanf( f, "\n" );

 return m;
}


void mmove_read( CamData  *cd, char *kfname, char *rtfname )
{
 int i;
 FILE *kfile, *rtfile;

 kfile = fopen( kfname, "r" );
 rtfile = fopen( rtfname, "r" );
 cd->k = kmatrix_read( kfile );
 while( !feof(rtfile) ){
   fscanf( rtfile, "Frame %d\n", &i );
   cd->rt[i] = rtmatrix_read(rtfile);
 }
 cd->nframes = i+1;
```

```
 fclose( kfile );
 fclose( rtfile );
}


void mmove_view( View *view, CamData *cd, int frame )
{
 Real mx, my, l, r, b, t;
 Matrix4 vi, v;
 Vector3 p, v1, v2, v3;
 view->V = cd->rt[frame];

 vi = m4_ident();
 vi.r1.x = v.r1.x ; vi.r1.y = v.r2.x ; vi.r1.z = v.r3.x ;
 vi.r2.x = v.r1.y ; vi.r2.y = v.r2.y ; vi.r2.z = v.r3.y ;
 vi.r3.x = v.r1.z ; vi.r3.y = v.r2.z ; vi.r3.z = v.r3.z ;
 p = v3_make( v.r1.w, v.r2.w, v.r3.w );
 v1 = v3_make( vi.r1.x, vi.r1.y, vi.r1.z );
 v2 = v3_make( vi.r2.x, vi.r2.y, vi.r2.z );
 v3 = v3_make( vi.r3.x, vi.r3.y, vi.r3.z );
 vi.r1.w = -v3_dot( v1 , p );
 vi.r2.w = -v3_dot( v2 , p );
 vi.r3.w = -v3_dot( v3 , p );
 view->Vinv = vi;

 mx =  cd->k.r1.x / cd->near;
 my =  cd->k.r2.y / cd->near;
 l =  -cd->k.r1.z/mx;
 r =  (cd->w - cd->k.r1.z)/mx;
 b =  -cd->k.r2.z/my;
 t =  (cd->h - cd->k.r2.z )/my;
 frustrum( view, l, b, r, t, cd->near, cd->far );
}
```

CHAPTER 3

# Optimization Tools

IN THIS CHAPTER we will talk about optimization, which consists on finding the minimum of a function over a specified domain.

These problems are fundamental in Computer Vision, since this field deals with estimation of models, and we want to make the best estimation for them. It is also necessary in the estimation of HDR Images, subject covered in the Chapter 10.

Firstly, we will present an optimization algorithm used for finding the minimum of a real function defined on an interval. In sequence we will present the Least-Squares Algorithm. After that we will explain the Levenberg-Marquardt algorithm, used for Non-linear least squares problems in $\mathbb{R}^n$. In these methods we will always talk about optimization without constraints.

In the second part of this chapter we will cover other important problem in Computer Vision, the solution of a homogeneous system of linear equations in a projective space, considering an over constrained situation. We will model it as a problem of minimizing the norm of the application of a linear function over a sphere.

Finally, we will present methods designed for solving optimize problems in the presence of noise. Those methods are fundamental, since the data used for estimating models in Computer Vision are very often corrupted by the presence of noise.

## 3.1 MINIMIZE A FUNCTION DEFINED ON AN INTERVAL

The most simple optimization problem of our interest consists on finding the minimal of a function over $\mathbb{R}$. More precisely, we will explain the golden section algorithm. This is a method for finding the minimum of an unimodal real function $f : [0, \infty] \to \mathbb{R}$.

$f$ is unimodal when the set of minimums $[t_1, t_2]$ is exclusively decreasing in $[0, t_1]$ and exclusively increasingly over the interval $[t_2, \infty]$

Consider that the minimum of $f$ is inside $[a, b] \subset [0, \infty]$. If we chose $u \in [a, b]$ and $v \in [a, b]$ such that $u < v$ then:

1. if $f(u) < f(v)$, then the interval $[v, b]$ cannot contain a minimum.

2. if $f(u) \geqslant f(v)$, then the interval $[a, u]$ cannot contain a minimum.

DOI: 10.1201/9781003206026-3

These conditions can be used for cutting the interval $[a, b]$ reducing it. After that, we can repeat the process many times to cut the generated intervals reducing them to a very small size.

Although $u$ and $v$ can be arbitrary chosen, an important question consists on defining a good choice for them. As can be viewed in [22], a good choice consists in using a method that makes use of the golden-section for it. This is called Golden Section algorithm for optimization.

In this method, we choose $u = a + \alpha(b - a)$ and $v = a + \beta(b - a)$ such that:

$$\alpha = \frac{3 - \sqrt{5}}{2}$$

and

$$\beta = \frac{\sqrt{5} - 1}{2}$$

At each step we discard $[a, u]$ or $[v, b]$ obtaining a new interval $[a', b']$ that will be partitioned by the points $u'$ and $v'$. Using the $\alpha$ and $\beta$ defined previously, we have that if $[v, b]$ is discarded then $v' = u$, and if $[a, u]$ is discarded then $u' = v$. It means that we only need to calculate one of the values $u'$ or $v'$ because the other value has already been calculated in the previous step. The Theorem 3.1 gives a proof of this fact. After applying many steps of this algorithm until find a sufficiently small segment $[a'', b'']$ with two inside points $u''$ and $v''$, then we define the minimum as

$$x = \frac{u'' + v''}{2}.$$

The GNU Scientific library uses an improved version of the Golden-Section algorithm known by Brent's minimization algorithm [6], which is the algorithm used for finding the minimum of real functions in the Chapter 7.

**Theorem 3.1.** *In the golden section algorithm, if $[v, b]$ is discarded then $v' = u$. In the same way, if $[a, u]$ is discarded then $u' = v$.*

**Proof**

Firstly we observe that using the definitions for $\alpha$ and $\beta$ we have that:

$$\alpha + \beta = 1, \tag{3.1}$$

$$\beta^2 = \alpha \tag{3.2}$$

$$1 - \alpha = \frac{\alpha}{1 - \alpha} \tag{3.3}$$

Considering that $[v, b]$ has been discarded then $b' = v$ and $a' = a$. Thus we have that

$$v' = a' + \beta(b' - a') = a + \beta(v - a) \tag{3.4}$$

By definition $v = a + \beta(b - a)$. Consequently

$$v - a = \beta(b - a) \tag{3.5}$$

Replacing it on Equation 3.4 we have that

$$v' = a + \beta^2(b - a). \tag{3.6}$$

Applying the Equation 3.2 we conclude that

$$v' = a + \alpha(b - a) = u \tag{3.7}$$

proving the first case.

Now, let us suppose that $[a, u]$ is discarded, thus $a' = u$ and $b' = b$. We have that

$$\begin{aligned} u' &= a' + \alpha(b' - a') = u + \alpha(b - u) = a + \alpha(b - a) + \alpha(b - (a + \alpha(b - a))) \\ &= a + (2\alpha - \alpha^2)(b - a). \end{aligned}$$

By the Equation 3.3 we obtain that $\alpha^2 = 3\alpha - 1$. Consequently

$$u' = a + (2\alpha - 3\alpha + 1)(b - a) = a + (1 - \alpha)(b - a) \tag{3.8}$$

Applying the Equation 3.1, we conclude that

$$u' = a + \beta(b - a) = v. \tag{3.9}$$

## 3.2 LEAST SQUARES

Let $A$ be a matrix $m \times n$ such that $m > n$, and $\mathbf{b} \in \mathbb{R}^n$. We want to find $\hat{\mathbf{x}} \in \mathbb{R}^n$ that minimizes $||A\mathbf{x} - \mathbf{b}||^2$.

Defining

$$A = \begin{pmatrix} a_{11} & a_{12} & \dots & a_{1n} \\ a_{21} & a_{22} & \dots & a_{2n} \\ \dots & \dots & \dots & \dots \\ a_{m1} & a_{m2} & \dots & a_{mn} \end{pmatrix}, \mathbf{x} = \begin{pmatrix} x_1 \\ x_2 \\ \dots \\ x_n \end{pmatrix} \text{ and } \mathbf{b} = \begin{pmatrix} b_1 \\ b_2 \\ \dots \\ b_n \end{pmatrix}$$

We have that

$$\min||A\mathbf{x} - \mathbf{b}||^2 = \min \sum_{i=1}^{m} (a_{i1}x_1 + \dots + a_{in}x_1 - b_i)^2,$$

which means that the least squares optimization is a good tool for solving over constrained linear systems.

Let $\hat{\mathbf{x}}$ be a solution to the problem. We have that $A\hat{\mathbf{x}}$ is the closest point to $\mathbf{b}$. As a consequence, $A\hat{\mathbf{x}} - \mathbf{b}$ must be orthogonal to the column space of $A$. In other words,

$$A^T(A\hat{\mathbf{x}} - \mathbf{b}) = 0. \tag{3.10}$$

Thus, we can find $\hat{\mathbf{x}}$ solving the linear system:

$$(A^T A)\mathbf{x} = A^T\mathbf{b}. \tag{3.11}$$

## 3.3 NON-LINEAR LEAST SQUARES

Another important tool in Computer Vision is the Non-linear least squares problems. Following we will present two algorithms for solving those problems: the Gauss-Newton method and the Levenberg-Marquardt algorithm.

### 3.3.1 Gauss-Newton Method

The Gauss-Newton method aims to find a minimum $\hat{x} \in \mathbb{R}^n$ for a function $g : \mathbb{R}^n \to R$ defined by $g(x) = \frac{1}{2}||f(x)||^2$, where $f : \mathbb{R}^n \to \mathbb{R}^m$ is a function defined so that close to $\hat{x}$ it is class $C^2$ . It uses the hypothesis that a point $\kappa_1 \in \mathbb{R}^n$ is known , which is an estimate for the minimum, or that is, $||\kappa_1 - \hat{x}||$ is small.

We can define the Taylor polynomial associated with $g$ at point $\kappa$ by

$$g(\kappa_1 + h) \approx g(\kappa_1) + g'(\kappa_1)u \cdot h + \frac{1}{2}g(\kappa_1)'' \cdot h \cdot h \tag{3.12}$$

As $g$ is differentiable, we have that $g$ assumes a minimum in $\kappa_1 + h$, if and only if, $g'(\kappa_1 + h) = 0$.

Using a first order approximation for $g$ we obtain

$$g'(\kappa_1 + h) = g''(\kappa_1) \cdot h + g'(\kappa_1). \tag{3.13}$$

Thus, in orther to find the vector $h$ that minimizes $g(\kappa_1 + h)$ we just have to solve the system

$$H_g(\kappa_1)h = -\nabla g(\kappa_1). \tag{3.14}$$

Using the fact that $g(x) = \frac{1}{2}||f(x)||^2$, we have that $g'(x) \cdot u = \langle f'(x) \cdot u, f(x) \rangle$, and consequently

$$g''(x) \cdot u \cdot v = \langle f''(x) \cdot u \cdot v, f(x) \rangle + \langle f'(x) \cdot u, f'(x) \cdot v \rangle \tag{3.15}$$

Using a first-order approximation for $f$, we obtain

$$g''(x) \cdot u \cdot v = \langle f'(x) \cdot u, f'(x) \cdot v \rangle. \tag{3.16}$$

We can rewrite using a matrix notation the relationships for the first and second derivatives obtained.

$$\nabla g(\kappa_1) = J_f(\kappa_1)f(\kappa_1), \tag{3.17}$$

and

$$Hg(\kappa_1) = J_f^T(\kappa_1)J_f(\kappa_1). \tag{3.18}$$

Replacing in the Equation 3.14, we have that $h$ can be estimated by the resolution of the system

$$J_f^T(\kappa_1)J_f(\kappa_1)h = -Jf(\kappa_1)f(\kappa_1). \tag{3.19}$$

Due to the approximations that are being made, we have in general that $\kappa_1 + h$ is not a minimum of $g$. What must be done is to define $\kappa_2 = \kappa_1 + h$ as a new estimate to the minimum, and the process is repeated until an $\kappa_i$ is obtained such that $\nabla g(\kappa_i)$ is considered small enough.

The convergence, or not, of the sequence $(\kappa_n)$ to $x$ will depend on the quality of the initial estimate $\kappa_1$. However, when this convergence occurs, it can be shown, as can be seen in [10], that it is of order two, that is, $\exists c \in \mathbb{R}$ such that

$$||\kappa_{i+1} - \hat{x}|| \leqslant c||\kappa_i - \hat{x}||^2. \tag{3.20}$$

### 3.3.2 Levenberg-Marquardt Algorithm

The Levenberg-Marquardt algorithm is an adaptation of the Gauss-Newton method used when the initial estimate for the minimum is not good enough to guarantee their convergence. The idea of the algorithm is to make a gradual transition from an optimization by descent by the gradient for the Gauss-Newton method, according to the estimation of the optimum point gets better and better. In the Levenberg-Marquadt algorithm we have that $\kappa_{i+1} = \kappa_i + h$, where $h$ is the solution of the system

$$(J_f^T(\kappa_i)J_f(\kappa_i) + \lambda I)h = -J_f(\kappa_i)f(\kappa_i). \tag{3.21}$$

We have that $\lambda \in \mathbb{R}$ is a value that can be modified at each iteration. $\lambda$ is initialized to a certain value, and at each iteration $\lambda$ can be multiplied or divided by a certain factor, in order to guarantee that the vector $h$ obtained produces a reduction in the objective function value.

The increase in the value of $\lambda$ causes the term $\lambda I$ to increase its importance when compared to $J_f^T(\kappa_i)J_f(\kappa_i)$. This makes the solution of the system 3.21 approximates to

$$h = -\lambda^{-1}J_f(\kappa_i)f(\kappa_i),$$

or, considering the Equation 3.17:

$$h = -\lambda^{-1}\nabla g(\kappa_i),$$

which makes the algorithm have a behavior similar to that of a gradient descent algorithm.

When $\kappa_i$ becomes closer to the optimal solution, the value of $\lambda$ decreases, causing the algorithm to behave similarly to the method Gauss-Newton, which accelerates convergence.

## 3.4 MINIMIZE THE NORM OF A LINEAR FUNCTION OVER A SPHERE

There is other important tool in Computer Vision: the solution of an homogeneous system of linear equations in a projective space, considering an over constrained situation. It can be solved in the following way:

Let $A$ be a $m \times n$ matrix. One way to reshape a system $Ax = 0$, with $x \in \mathbb{R}P^n$, in the case where the $m \geqslant n$, is to consider it as a solution for the problem of finding $x \in S^n$ that minimizes $||Ax||$. We will denote this problem as $min_{||x||=1}||Ax||$.

Its solution can be easily determined by the theorem below.

**Theorem 3.2.** *Let $Udiag(\lambda_1, \lambda_2, ..., , \lambda_2)V^T$ , with $\lambda_1 \geqslant \lambda_2 \geqslant ... \geqslant \lambda_n \geqslant 0$, the SVD decomposition of a matrix $A$, $m \times n$, such that $m \geqslant n$. If $v \in \mathbb{R}^n$ is the vector* *corresponding to the $n^{th}$ column of $V$, we have that $v$ is the vector that minimizes the function $x \mapsto Ax$, defined on the points of $\mathbb{R}^n$ that satisfy $\|x\| = 1$.*

**Proof**

$$min_{\|x\|=1}\|Ax\| = min_{\|x\|=1}\|USV^Tx\|, \tag{3.22}$$

such that $S = diag(\lambda_1, \lambda_2, ..., \lambda_n)$.

Since $U$ is an isometry

$$min_{\|x\|=1}\|USV^Tx\| = min_{\|x\|=1}\|SV^Tx\|. \tag{3.23}$$

Defining $y = V^Tx$, we get

$$min_{\|x\|=1}\|SV^Tx\| = min_{\|y\|=1}\|Sy\|. \tag{3.24}$$

Using the definition of $S$

$$min_{\|y\|=1}\|Sy\| = min_{y_1^2+...+y_n^2=1}\sqrt{\lambda_1^2y_1^2 + ...\lambda_n^2y_n^2} \tag{3.25}$$

As $\lambda_1 \geqslant \lambda_2 \geqslant ... \geqslant \lambda_n \geqslant 0$ we conclude that the solution to this problem of optimization is the vector $y = (0, ..., 0, 1)^T$ . Thus, the vector $v$ that solves $min_{\|x\|=1}\|Ax\|$ is given by $V(0, ..., 0, 1)^T$, which corresponds to the $n^{th}$ column of $V$.

## 3.5 TWO STAGES OPTIMIZATION

The Levenberg-Marquardt is an algorithm that is good for describing optimizations problems based on geometric errors for estimating models that adjust a set of points. The problem is that it needs an initial solution to be used. On the other hand many times it is possible to define optimization problems that can be computed without an approximation to the initial solution, whose theoretical optimum is the same as the one computed by the Levenbeg-Marquard algorithm, but minimizing considering an algebraic error without geometric interpretation.

As a consequence, in Computer Vision it is very common to couple an algorithm that find an initial solution considering an algebraic error with a second algorithm that solve the problem using the Levenberg-Marquardt solution, finding a response whose errors have geometric interpretations (Figure 3.1). One very common algorithm used for finding the initial solution is the one presented in the Section 3.4.

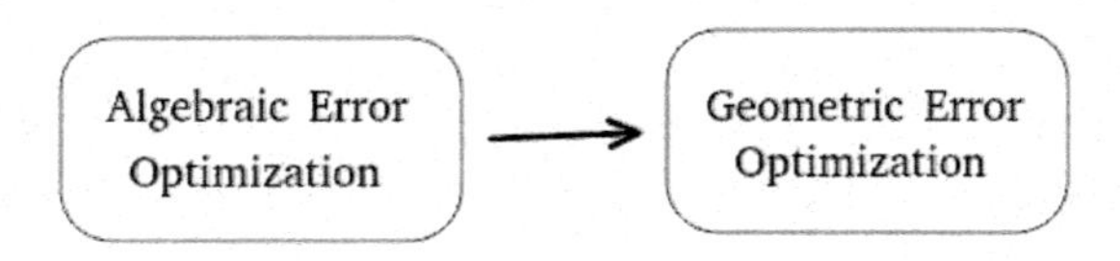


Figure 3.1 Coupling a optimization problem that do not need an initial solution, but whose error does not have geometric interpretation, with an optimization algorithm that finds an optimum with geometric interpretation.

## 3.6 ROBUST MODEL ESTIMATION

It is very common, in Computer Vision, situations where the observations used for estimating models are corrupted by noise. This noise defines lots of outliers that must be discarded before the optimization algorithm finds the estimated model. One way to deal with this situation is using the RANSAC algorithm.

### 3.6.1 RANSAC Algorithm

The RANSAC (Random Sample Consensus) algorithm was proposed by Fischler and Bolles in [8], where it was presented in the following terms:

"Given a model that requires a minimum of $n$ data points to instantiate its free parameters, and a set of data points $P$ such that the number of points in $P$ is greater than $n$ [$\sharp(P) \geqslant n$], randomly select a subset S1 of $n$ data points from $P$ and instantiate the model. Use the instantiated model $M1$ to determine the subset $S1*$ of points in $P$ that are within some error tolerance of $M1$. The set $S1*$ is called the consensus set of $S1$.

If $\sharp(S1*)$ is greater than some threshold $t$, which is a function of the estimate of the number of gross errors in $P$, use $S1*$ to compute (possibly using least squares) a new model $M1*$.

If $\sharp(S1*)$ is less than $t$, randomly select a new subset $S2$ and repeat the above process. If, after some predetermined number of trials, no consensus set with t or more members has been found, either solve the model with the largest consensus set found, or terminate in failure."

### 3.6.2 Example of Using the RANSAC Algorithm

Now we explain how to find a straight line in $\mathbb{R}^2$, defined by a set of sampled points such that some of them are corrupted by noise, defining outliers to the problem (Figure 3.2).

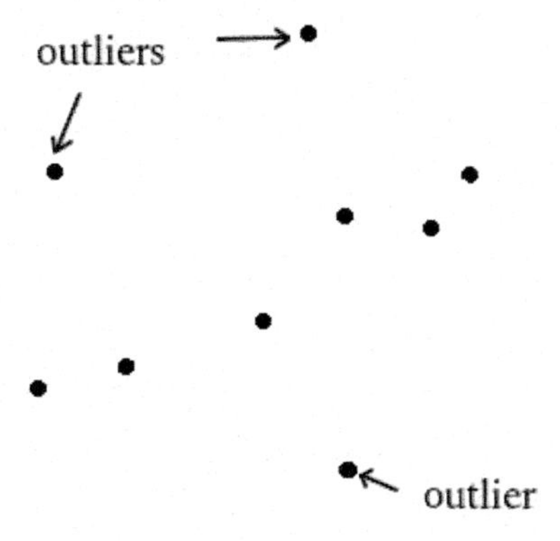


Figure 3.2 A set of points, corrupted by noise, defining a straight line.

We know that a line can be specified by two points. Thus, we select two random points $p$ and $q$ in the set of observed points and we count the number of points that are inside a tubular neighborhood of the line.

There are two different possible situations. One of them occurs when one of the selected points is an outlier. In this case, we expect that there are a few points in a tubular neighborhood of the defined line (Figure 3.3).

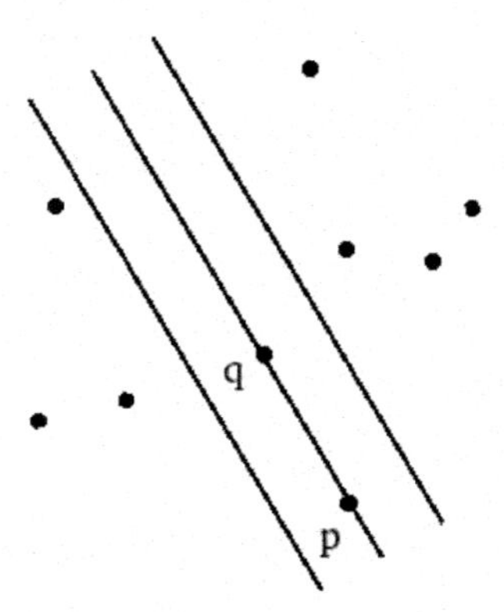


Figure 3.3 Selecting two points defining a straight line. In this case one of the points is an outlier, and there are few points inside a tubular neighborhood of the line.

The second possibility occurs when both selected points are inliers (Figure 3.4). In this case, there are many points in a tubular neighborhood of the defined line. As a consequence, there are many points inside the tubular neighborhood around the line defined by $p$ and $q$.

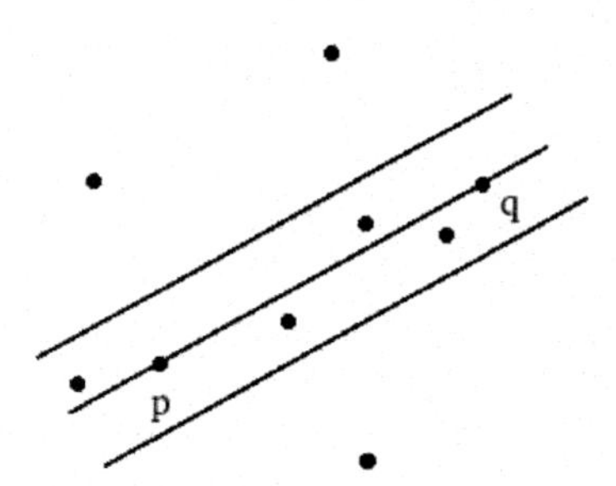


Figure 3.4 Selecting two points defining a straight line. In this case both points are inliers, and there are many points inside a tubular neighborhood of the line.

After choosing the straight line with the largest number of observed points in a tubular neighborhood of it, the final estimated line can be computed by a non-linear optimization, minimizing the distance of the inliers points to it.

CHAPTER 4

# Estimating One Camera

IN THIS CHAPTER we will talk about the estimating of one camera. Initially we will explain how to do it based on a observation of a set of a 3D points. After that we will explain how to factorize the estimated camera in the form $K[R|t]$. After that we will explain how these extrinsic parameters can be adjusted to be used in Computer Graphics system. And in the end of this chapter, we will explain how to estimate a camera using a planar pattern.

## 4.1 CALIBRATION IN RELATION TO A SET OF 3D POINTS

The calibration problem in relation to a set of 3D points can be defined by:

**Problem 4.1.1.** *Knowing the projections* $x_1, ..., x_n$*, with* $x_i \in \mathbb{R}P^2$*, corresponding to points* $X_1, ..., X_n$*, with* $X_i \in \mathbb{R}P^3$*. Determine the transformation* $P : \mathbb{R}P^3 \to \mathbb{R}P^2$*, such that* $PX_i = x_i, i \in \{1, 2, ..., n\}$*.*

### 4.1.1 Calibration Using Six Matches

Considering the elements of the matrix associated with $P$ as variables, each sentence of the form $PX_i = x_i$ defines two linear equations with 12 variables. Consequentially, if 6 correspondences between points and projections are established, it is known that the system has a solution if there are no linearly dependent lines.

As the coordinates of each $x_i$ are normally corrupted by noise, for example, obtained by a camera, an error is introduced in the system solution, making interesting to use a larger number of correspondences in a super determined formulation. In addition, the system obtained with 6 matches, although it may seem well determined, it is not. It is a super-determined system, since $P$ is defined less than a scalar multiplication.

We will show below how this problem can be solved using an arbitrary amount of correspondence.

DOI: 10.1201/9781003206026-4

### 4.1.2 Calibration Using More Than Six Matches

In order to solve the calibration problem using more than six matches, we just have to consider that finding $P$ that satisfies

$$\forall i \in 1, ..., n, PX_i = (u_i, v_i, 1)^T \tag{4.1}$$

is equivalent to solving the system $AP = 0$, where [13]

$$A = \begin{pmatrix} X_1^T & 0^T & -u_1 X^T 1 \\ 0^T & X_1^T & -v_1 X_1^T \\ X_2^T & 0^T & -u_2 X_2^T \\ 0^T & X_2^T & -v_2 X_2^T \\ \vdots & \vdots & \vdots \\ X_n^T & 0^T & -u_n X_n^T \\ 0^T & X_n^T & -v_n X_n^T \end{pmatrix}, \tag{4.2}$$

and $P = (P_{11}, P_{12}, ..., P_{33}, P_{34})^T$ is a vector whose elements are the 12 elements of the matrix $P$, to be determined.

We can use Theorem 3.2 to solve the problem $min_{||P||=1}||AP||$, which provides an estimate for the elements of the $P$ matrix.

## 4.2 NORMALIZATION OF THE POINTS

The algorithm of the previous section is poorly conditioned. [13] describes a better conditioned version of it consists on applying a homography to the 2D points and other to the 3D points before applying the calibration process. After that the inverse of the homographies must be used to denormalize the camera matrix.

More precisely, the homography $H$ that must be applied to the points must make the centroid of the points in the origin and scale the points in such a way that the root-mean-square of the distances between the points to the origin is $\sqrt{2}$. The 3D homography $T$ that must be applied to the 3D points must make the centroid also becomes the origin and makes the root-mean-square of the distances between the points to the origin equals to $\sqrt{3}$.

After making this change to the 2D and to the 3D points we must apply the algorithm described in Section 4.1.2 finding the camera $P$. The camera found by this process is $H^{-1}PT$.

## 4.3 ISOLATION OF CAMERA PARAMETERS

Let us consider that we are in possession of a matrix $P$, $3 \times 4$, which represents a projective transformation. We will now show a process for factoring $P$ in the form $K[R|t]$. This process is important for two reasons. On the one hand, it functions as a demonstration, by construction, that projective transformations defined in $\mathbb{R}P^3 \rightarrow \mathbb{R}P^2$ are always models for generic projective cameras. On the other hand, it serves as a algorithm to determine the intrinsic and extrinsic parameters of a camera.

Let $P = \lambda K[R|t]$, where $\lambda$ is a constant that can take any value in $\mathbb{R} - \{0\}$. Assuming the following definitions:

$$P = \begin{pmatrix} a_1^T & b_1 \\ a_2^T & b_2 \\ a_3^T & b_3 \end{pmatrix}, K = \begin{pmatrix} f_1 & s & u_0 \\ 0 & f_2 & v_0 \\ 0 & 0 & 1 \end{pmatrix} and [R|t] = \begin{pmatrix} R_1^T & t_1 \\ R_2^T & t_2 \\ R_3^T & t_3 \end{pmatrix}.$$

We have that

$$P = \begin{pmatrix} a_1^T & b_1 \\ a_2^T & b_2 \\ a_3^T & b_3 \end{pmatrix} = \lambda \begin{pmatrix} f_1 R_1^T + sR_2^T + u_0 R_3^T & f_1 t_1 + st_2 + u_0 t_3 \\ f_2 R_2^T + v_0 R_3^T & f_2 t_2 + v_0 t_3 \\ R_3^T & t_3 \end{pmatrix}$$

We will now show you how to determine all intrinsic and extrinsic parameters associated with $P$.

**Determining $\lambda$**

We can determine $|\lambda|$ using that $|\lambda| = |\lambda|||R_3^T|| = ||a_3^T||$. We will assume for while that $\lambda > 0$, in the end we will conclude whether or not this choice was appropriate. Or that is, let's assume that

$$\lambda = ||a_3^T||. \tag{4.3}$$

**Determining $R_3$ and $t_3$**

Defining $P' = \frac{1}{\lambda}P$ we get

$$P' = K[R|t] = \begin{pmatrix} a_1'^T & b_1' \\ a_2'^T & b_2' \\ a_3'^T & b_3' \end{pmatrix} = \begin{pmatrix} f_1 R_1^T + sR_2^T + u_0 R_3^T & f_1 t_1 + st_2 + u_0 t_3 \\ f_2 R_2^T + v_0 R_3^T & f_2 t_2 + v_0 t_3 \\ R_3^T & t_3 \end{pmatrix}.$$

As a consequence we have that

$$R_3 = a_3', \tag{4.4}$$

and

$$t_3 = b_3'. \tag{4.5}$$

**Determining $v_0$**

To determine $v_0$ it is enough to observe that

$$f_2 R_2^T + v_0 R_3^T = a_2'^T \Rightarrow f_2 R_2^T R_3 + v_0 R_3^T R_3 = a_2'^T R_3 \tag{4.6}$$

Like $R_2 \perp R_3$ and $R_3^T R_3 = 1$ we conclude that

$$v_0 = a_2'^T R_3. \tag{4.7}$$

**Determining $u_0$**

To determine $u_0$, just note that

$$f_1 R_1^T + sR_2^T + u_0 R_3^T = a_1'^T \Rightarrow f_1 R_1^T R_3 + sR_2^T R_3 + u_0 R_3^T R_3 = a_1'^T R_3. \tag{4.8}$$

Since $R_1 \perp R_3$ , $R_2 \perp R_3$ and $R_3^T R_3 = 1$ we conclude that

$$u_0 = a_1'^T R_3. \tag{4.9}$$

**Determining** $f_2$, $R_2$ and $t_2$

We have that

$$f_2 R_2^T + v_0 R_3^T = a_2'^T \Rightarrow f_2 R_2^T = a_2'^T - v_0 R_3^T \Rightarrow ||f_2|| = ||a_2'^T - v_0 R_3^T||.$$

Since the sing of $f_2$ is positive, we conclude that

$$f_2 = ||a_2^T - v_0 R_3^T||. \tag{4.10}$$

To determine $R_2$ and $t_2$ we use that

$$R_2^T = \frac{1}{f_2}(a_2'^T - v_0 R_3^T) \tag{4.11}$$

and

$$t_2 = \frac{1}{f_2}(b_2' - v_0 t_3). \tag{4.12}$$

**Determining** $R_1$

$R_1$ can be obtained directly from $R_2$ and $R_3$ considering that $R$ is a rotation. In other words

$$R_1 = R_2 \times R_3. \tag{4.13}$$

**Determining** $f_1$, $s$ and $t_1$. We have that

$$f_1 R_1^T + s R_2^T + u_0 R_3^T = a_1'^T \Rightarrow f_1 R_1^T R_2 + s R_2^T R_2 + u_0 R_3^T R_2 = a_1'^T R_2 \tag{4.14}$$

Since $R_1 \perp R_2$, $R_2 \perp R_3$ and $R_2^T R_2 = 1$, we have that

$$s = a_1'^T R_2. \tag{4.15}$$

Based on a similar reasoning, we can conclude that

$$f_1 = a_1'^T R_1. \tag{4.16}$$

The value of $t_1$ can be obtained by observing that

$$t_1 = \frac{1}{f_1}(b_1' - u_0 t_3 - s t_2). \tag{4.17}$$

**Correcting the** $\lambda$

We are interested in defining a coordinate system associated with the camera that satisfies $i \times j = k$, where $k$ specifies the view direction of the camera.

The procedure for calculating intrinsic parameters previously described a solution that satisfies $f_2 > 0$, but the sign of $f_1$ can be both positive and negative.

If we want the image coordinate system to be defined with origin in the lower left corner (Figure 4.1-a), we need to have $f_2 > 0$ and $f_1 < 0$. On the other hand,

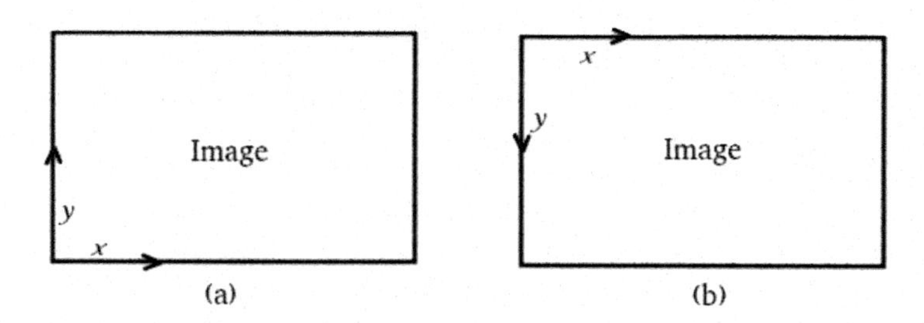


Figure 4.1 (a) An image with the coordinate system in the lower left corner. (b) An image with the coordinate system in the top left corner.

if we want a coordinate system defined with the origin in the top left corner (Figure 4.1-b), we need to have $f_2 > 0$ and $f_1 > 0$.

After calculating the previous parameters we can correct the sign of $f_1$ applying the following transformation on the answer found if necessary:

$$\begin{pmatrix} f_1 & s & u_0 \\ 0 & f_2 & v_0 \\ 0 & 0 & 1 \end{pmatrix} \begin{pmatrix} R_1^T & t_1 \\ R_2^T & t_2 \\ R_3^T & t_3 \end{pmatrix} \mapsto \begin{pmatrix} -f_1 & s_u & u_0 \\ 0 & f_2 & v_0 \\ 0 & 0 & 1 \end{pmatrix} \begin{pmatrix} R_1^T & t_1 \\ -R_2^T & -t_2 \\ -R_3^T & -t_3 \end{pmatrix}.$$

This transform do not change $P$, because of the Theorem 1.1, since it can be interpreted as a change in the sign of $\lambda$.

## 4.4 CAMERA FOR IMAGE SYNTHESIS

The intrinsic parameters obtained by the factoring of a generic projective camera present a degree of freedom that does not exist in the camera models used in synthesis of images. If the matrix of a camera's intrinsic parameters is

$$K = \begin{pmatrix} f_1 & s & u_0 \\ 0 & f_2 & v_0 \\ 0 & 0 & 1 \end{pmatrix} \tag{4.18}$$

it can only be framed in the traditional models of image synthesis in the case that $s = 0$.

Since we estimate the generic projective camera from measurements made by a camera that was manufactured with the purpose of presenting the property $s = 0$, we have that the obtained matrix $K$ will be defined with a small value for $|s|$. Because of this, the effect of the transformation $K'$ obtained by substituting the value of $s$ for zero in $K$, is similar to that of the $K$ transformation, whose difference is that the first can be adapted to the purpose of image synthesis.

We will not use the $K'$ matrix directly, we will show how to use it as a starting point for an optimization algorithm that you will find the solution we are looking for.

## 4.5 CALIBRATION BY RESTRICTED OPTIMIZATION

Consider the following problem:

**Problem 4.5.1.** *Let $\Omega$ be the space of the projective cameras such that their matrices of intrinsic parameters satisfy the constraint $s = 0$ following the notation of the Equation 4.18. Known the projections $x_1, \dots, x_n$, with $x_i \in \mathbb{R}P^2$ , corresponding to the points $X_1, ..., X_n$ , with $X_i \in \mathbb{R}P^3$. Determine the transformation $P = K[R|t] \in \Omega$, such that $\sum_{i=0}^{n} d(PX_i, x_i)^2$ is minimal.*

This formulation for the camera calibration problem presents two important aspects:

1. The objective function has a geometric meaning based on the reprojection error of points $X_1, \dots, X_n$, which is more natural than the algebraic error defined in the Section 4.1.2.
2. The solution found is optimal in space $\Omega$, that is, the matrix $K$ is the best choice of intrinsic parameters that can be made to explain the projections of $X_1, \dots, X_n$, maintaining compatibility with the model employed in image synthesis.

The Levenberg-Marquardt optimization algorithm can be used to solve this problem. To use it, it is necessary that an element of $\Omega$ is known close to the optimal solution. A camera with this property can be found as follows manner:

1. A generic projective camera $P$ is estimated, as described in Section 4.1.
2. Factor $P$ into the form $K[R|t]$, as described in Section 4.3.
3. The intrinsic parameter $s$ of the matrix $K$ is replaced by zero, as described in Section 4.4, thus obtaining the $K[R|t] \in \Omega$ camera.

### 4.5.1 Adjusting the Levenberg-Marquardt to the Problem

We can employ the Levenberg-Marquardt algorithm to solve the Problem 4.5.1. For this, using the notation established in this problem, we will define a function $\psi : U \subset \mathbb{R}^{10} \to (\mathbb{R}^2)^m$ by

$$\psi_i(z) = P(z)X_i, \tag{4.19}$$

for $i \in \{1, \dots, m\}$, where $U$ and $P : U \to \Omega$ are defined so that $P$ is an onto function in the subset of $\Omega$ of the cameras that applied to $X_1, \dots, X_m$ generate projections that are affine points of $\mathbb{R}P^2$.

In order to solve the Problem 4.5.1 we can use the Levenberg-Marquardt algorithm to find the minimum of the function $g : U \to \mathbb{R}$ defined by

$$g(z) = \frac{1}{2}||\psi(z) - (x_1, \dots, x_m)^T||^2, \tag{4.20}$$

where $(x_1, \dots, x_m)$ are the observed 2D projections.

We emphasize that when we consider the image of each $\psi_i$ as a vector of $\mathbb{R}^2$ , we are embedding the transformation defined by $(x, y, z)^T \mapsto (\frac{x}{z}, \frac{y}{z})^T$, that makes the conversion of coordinates from related points of $\mathbb{R}P^2$ to coordinates of $\mathbb{R}^2$, and we are doing the same with the homogeneous coordinates of $x_1, \ldots, x_m$ . We will present in Section 4.5.3 a definition of $P$ that makes the function of class $C^2$ for almost all $U$ points. Enabling the use of the Levenberg-Marquardt algorithm. Before that, we will show how to obtain a parameterization for a rotation via specification of an axis and a rotation angle.

### 4.5.2 Parameterization of Rotations

The angle rotation $\theta \in \mathbb{R}$, around an axis specified by the vector $\omega = (\omega_1, \omega_2, \omega_3)$ $T \in \mathbb{R}^3$ , with $||\omega|| = 1$, is given by the linear transformation $\mathbb{R} : \mathbb{R}^3 \to \mathbb{R}^3$ whose matrix representation is [7]:

$$R = \begin{pmatrix} \omega_1^2 + C(1-\omega_1^2) & \omega_1\omega_2(1-C) - \omega_3 S & \omega_1\omega_3(1-C) + \omega_2 S \\ \omega_1\omega_2(1-C) + \omega_3 S & \omega_2^2 + C(1-\omega_2^2) & \omega_2\omega_3(1-C) - \omega_1 S \\ \omega_1\omega_3(1-C) - \omega_2 S & \omega_2\omega_3(1-C) + \omega_1 S & \omega_3^2 + C(1-\omega_3^2) \end{pmatrix},$$

where $C = cos\theta$, and $S = \sin\theta$.

Conversely, the axis of rotation $\omega$ and angle $\theta$ can be obtained from the transformation $R$. For that, it is enough to observe that the sub-space generated by this axis is invariant by $R$, that is, $\omega$ is a eigenvector of $R$. Furthermore, the restriction of $R$ to subspace generated by $\omega$ is the identity transformation, therefore the eigenvalue associated with $\omega$ it is unitary. That is, to determine $\omega$, it is enough to find a non-trivial solution for the equation $(R - I)\omega = 0$, which can be obtained by Theorem 3.2 The angle $\theta$, can be determined using that

$$cos\theta = \frac{1}{2}(tr(R) - 1), \tag{4.21}$$

$$\sin\theta = \frac{1}{2}\langle\omega, \tau\rangle, \tag{4.22}$$

such that $\tau = (R_{32} - R_{23}, R_{13} - R_{31}, R_{21} - R_{12})^T$.

To obtain an explicit representation for the three degrees of freedom associated associated with a rotation, just use the vector $\omega' \in \mathbb{R}^3$ defined by $\omega' = \theta\omega$.

Obtaining $\omega$ and $\theta$ from $\omega'$ is done as follows

1. If $||\omega'|| \neq 0$, then $\omega = \frac{\omega'}{||\omega'||}$ and $\theta = ||\omega'||$.

2. If $||\omega'|| = 0$, then $\theta = 0$ and $\omega$ can be any unit vector.

### 4.5.3 Parameterization of the Camera Space

Using the parameterization of the rotation matrices presented above we can define the function $P : U \subset \mathbb{R}^{10} \to \Omega$, used to define $\psi$ in Equation 4.19, satisfying the properties required in Section 4.5.1

$$P(f_1, f_2, u_0, v_0, t_1, t_2, t_3, \omega_1, \omega_2, \omega_3) = \begin{pmatrix} f_1 R_1^T(\omega) + u_0 R_3^T(\omega) & f_1 t_1 + u_0 t_3 \\ f_2 R_2^T(\omega) + v_0 R_3^T(\omega) & f_2 t_2 + v_0 t_3 \\ R_3^T(\omega) & t_3 \end{pmatrix},$$

such that

$$\begin{pmatrix} R_1^T(\omega) \\ R_2^T(\omega) \\ R_3^T(\omega) \end{pmatrix}$$

is the matrix that, for $||\omega|| \neq 0$, represents a rotation of $||\omega||$ radians around the axis $\omega = (\omega_1, \omega_2, \omega_3)^T$. And for $||\omega|| = 0$ is the matrix identity.

Since $\forall z \in U$, $P(z)X_i$ is an affine point of $\mathbb{R}P^2$, we have that $\psi_i(z)$ is being defined by

$$\left( \frac{f_1 R_1^T(\omega)X_i + u_0 R_3^T(\omega)X_i + f_1 t_1 + u_0 t_3}{R_3^T(\omega)X_i + t_3}, \frac{f_2 R_2^T(\omega)X_i + u_0 R_3^T(\omega)X_i + f_2 t_2 + v_0 t_3}{R_3^T(\omega)X_i + t_3} \right)^T$$

## 4.6 PROBLEM POINTS OF PARAMETERIZATION

The Levenberg-Marquardt algorithm evaluates $P$ in a sequence of elements of the euclidean space $\mathbb{R}^{10}$.

Cameras that are not parameterized by $U$ points are those that satisfy of $R_3^T(\omega)X_i + t_3 = 0$, for some $i \in \{1, \ldots, m\}$. These are the configurations that cause any of the $X_i$ to have no projection, which occurs when the $z$ coordinate of $X_i$ is null in the camera reference. In order to see this just remember that $t_3 = -R_3^T c$, in which $c$ is the position of the projection center, and note that

$$R_3^T(\omega)X_i + t_3 = 0 \Leftrightarrow R_3^T(\omega)(X_i - c) = 0 \Leftrightarrow [R(\omega)(X_i - c)]_3 = 0.$$

This region of $\mathbb{R}^{10}$ in which $\psi$ is not defined, nor its derivatives, does not generate problems during the execution of the Levenberg-Marquardt algorithm. The reason for this to happen is the fact that the objective function assumes very high values at $U$ points that belong to small neighborhoods of these configurations, because the reprojection error associated with any of the $X_i$ is very large. Consequently, the configuration sequences generated by the iterations of the algorithm they should not approach that region.

## 4.7 FINDING THE INTRINSIC PARAMETERS

We can use the calibration process described in the previous sections to find the intrinsic parameters of a camera. In order to do this, we need a 3D calibration pattern with markers in known places (Figure 4.2).

Each marker can be placed in correspondence with their respective projection in the captured image allowing us to perform the calibration.

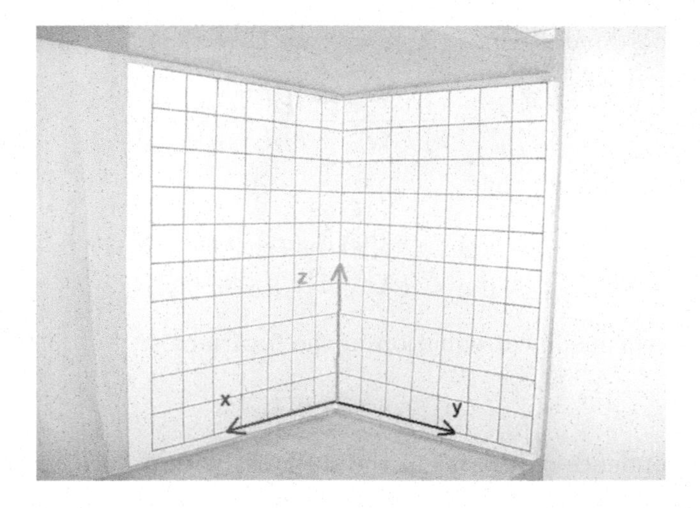


Figure 4.2 3D Pattern used for estimating the intrinsic parameters of a camera, and its coordinate system.

## 4.8 CALIBRATION USING A PLANAR PATTERN

Although mostly of this book cover the MatchMove process, we decided to include this section describing a calibration method based on a planar pattern. We did this because sometimes we are interested in visual effects in which the camera is not moving and the scene is over a plane. Besides that, the following method can also be used to find the intrinsic parameter of a camera.

This section presents the camera calibration algorithm proposed by Roger Y. Tsai in [30], and revisited in [15]. Here we will not address issues related to distortion modeling generated by the use of lenses, which is something possible with the model proposed by Tsai. We will assume, such as in the rest of this work, that these distortions can be overlooked. In addition, we will only deal with the case where the calibration pattern is flat.

Let $n$ points of a flat calibration pattern $X_1, \ldots, X_n$ be known, with $X_i \in \mathbb{R}^3$, such that $(X_i)_z = 0$, and their respective projections are also known $x_1, \ldots, x_n$, with $x_i \in \mathbb{R}^2$ obtained by a camera. We want to get a $K[R|t]$ model for this camera.

Let's assume that the principal point $(x_0, y_0)^T$ of the camera is known. This fact it is not absurd, because normally this point is close to the central pixel of the Image. Let us define the set $x'_1, \ldots, x'_n$, obtained from the projections, considering that:

$$x'_i = x_i - (x_0, y_0)^T \tag{4.23}$$

Let's assume that

$$K = \begin{pmatrix} f & 0 & x_0 \\ 0 & f & y_0 \\ 0 & 0 & 1 \end{pmatrix} \quad [R|t] = \begin{pmatrix} r_{11}r_{12}r_{13}t_1 \\ r_{21}r_{22}r_{23}t_2 \\ r_{31}r_{32}r_{33}t_3 \end{pmatrix} \tag{4.24}$$

We have, for all $i \in 1, ..., n$ that

$$x_i = \lambda K[R|t]X_i. \tag{4.25}$$

In terms of coordinates, since $(X_i)_z = 0$, we have

$$\frac{(x'_i)_x}{(x'_i)_y} = \frac{r_{11}(X_i)_x + r_{12}(X_i)_y + t_1}{r_{21}(X_i)_x + r_{22}(X_i)_y + t_2}. \tag{4.26}$$

Consequently, it follows that

$$\begin{gathered}[(X_i)_x(x'_i)_y]r_{11} + [(X_i)_y(x'_i)_y]r_{12} + (x'_i)_y t_1 - \\ [(X_i)_x(x'_i)_x]r_{21} + [(X_i)_y(x'_i)_x]r_{22} + (x'_i)_x t_2 = 0,\end{gathered} \tag{4.27}$$

which is a homogeneous linear equation in the 6 variables:

$$r_{11}, r_{12}, r_{21}, r_{22}, t_1, t_2.$$

Each correspondence of a point in the scene $X_i$ with its projection $x_i$ provides a homogeneous equation. So, if 6 or more matches are established you can find these parameters, less than a scale factor, using Theorem 3.2.

Let us suppose that the solution found was

$$\hat{r}_{11}, \hat{r}_{12}, \hat{r}_{21}, \hat{r}_{22}, \hat{t}_1 \text{ e } \hat{t}_2.$$

We have that $\exists \lambda \in \mathbb{R}$ such that

$$\hat{r}_{11}^2 + \hat{r}_{12}^2 + \hat{r}_{13}^2 = \lambda^2, \tag{4.28}$$

$$\hat{r}_{21}^2 + \hat{r}_{22}^2 + \hat{r}_{23}^2 = \lambda^2, \tag{4.29}$$

$$\hat{r}_{11}\hat{r}_{21} + \hat{r}_{12}\hat{r}_{22} + \hat{r}_{13}\hat{r}_{23} = 0. \tag{4.30}$$

For the first two equations we have

$$(\hat{r}_{11}^2 + \hat{r}_{12}^2)(\hat{r}_{21}^2 + \hat{r}_{22}^2) = (\lambda^2 - \hat{r}_{13}^2)(\lambda^2 - \hat{r}_{23}^2). \tag{4.31}$$

By the third equation we have

$$(\hat{r}_{11}\hat{r}_{21} + \hat{r}_{12}\hat{r}_{22})^2 = (\hat{r}_{13}\hat{r}_{23})^2. \tag{4.32}$$

Subtracting the Equation 4.32 from the Equation 4.31 we get that

$$(\hat{r}_{11}\hat{r}_{22} - \hat{r}_{12}\hat{r}_{21})^2 = \lambda^4 - \lambda^2(\hat{r}_{13}^2 + \hat{r}_{23}^2). \tag{4.33}$$

By the Equations 4.28 and 4.29 we have that

$$(\hat{r}_{13}^2 + \hat{r}_{23}^2) = 2\lambda^2 - (\hat{r}_{11}^2 + \hat{r}_{12}^2 + \hat{r}_{21}^2 + \hat{r}_{22}^2) \tag{4.34}$$

we conclude that

$$\lambda^4 - \lambda^2(\hat{r}_{11}^2 + \hat{r}_{12}^2 + \hat{r}_{21}^2 + \hat{r}_{22}^2) + (r'_{11}\hat{r}_{22} - \hat{r}_{12}\hat{r}_{21})^2 = 0, \tag{4.35}$$

which is a second degree equation in $\lambda^2$. Solving this equation we can calculate $\hat{r}_{13}$ and $\hat{r}_{23}$ using that

$$\hat{r}_{13}^2 = \lambda^2 - (\hat{r}_{11}^2 + \hat{r}_{12}^2) \tag{4.36}$$

and

$$\hat{r}_{23}^2 = \lambda^2 - (\hat{r}_{21}^2 + \hat{r}_{22}^2). \tag{4.37}$$

We have that only the biggest root of equation 4.35 makes the right sides of these equalities are positive. so we just need to consider that root.

We can then calculate the first two lines of the rotation matrix by dividing $\hat{r}_{11}, \hat{r}_{12}, \hat{r}_{13}, \hat{r}_{21}, \hat{r}_{22}$ and $\hat{r}_{23}$ for $\lambda$. After that, we can find the third row by taking the cross product of the first two. We have, however, that there is a sign ambiguity in the calculation of $\hat{r}_{13}$ and $\hat{r}_{23}$ obtained by equations 4.36 and 4.37.

We can find the signal of $\hat{r}_{13}\hat{r}_{23}$ using, as a consequence of Equation 4.30, that

$$\hat{r}_{13}\hat{r}_{23} = -(\hat{r_{11}}\hat{r_{21}} + \hat{r_{12}}\hat{r_{22}}). \tag{4.38}$$

To determine the signs of $\hat{r}_{13}$ and $\hat{r}_{23}$ we arbitrarily choose one of the choices possible signals. If the choice of the signal is correct, we have to have the projected image of a point of the calibration pattern obtained by applying the resulting camera model must be close to the projections actually obtained with the real camera. More precisely, where $x_p$ is the image predicted by the model and $x_o$ is the image actually obtained by the camera real, considering a coordinate system in which it is assigned to the principal point the coordinates $(0,0)^T$, we have to choose the signs for $\hat{r}_{13}$ and $\hat{r}_{23}$ that make [15]:

$$\langle x_p, x_o \rangle \geqslant 0. \tag{4.39}$$

If the choice was incorrect, just change the signs of $\hat{r}_{13}$ or $\hat{r}_{23}$.

We still need to determine $f$ and $t_3$. These parameters can be estimated using that

$$\frac{(x_i')_x}{f} = \frac{r_{11}(X_i)_x + r_{12}(X_i)_y + t_1}{r_{31}(X_i)_x + r_{32}(X_i)_y + t_3} \tag{4.40}$$

and

$$\frac{(x_i')_x}{f} = \frac{r_{21}(X_i)_x + r_{22}(X_i)_y + t_2}{r_{31}(X_i)_x + r_{32}(X_i)_y + t_3}. \tag{4.41}$$

Performing the cross multiplication we obtain the following system of linear equations, that allows us to calculate $f$ and $t_3$, taking into account that all the others parameters that appeared have been estimated:

$$\begin{cases} (r_{11}(X_i)_x + r_{12}(X_i)_y + t_1)f - (x_i')_x t_3 = (r_{31}(X_i)_x + r_{32}(X_i)_y)(x_i')_x \\ (r_{21}(X_i)_x + r_{22}(X_i)_y + t_2)f - (x_i')_y t_3 = (r_{31}(X_i)_x + r_{32}(X_i)_y)(x_i')_y \end{cases}$$

Once all the extrinsic parameters and the $f$ parameter have been estimated, we can perform a refinement of the solution by applying the Levenberg-Marquardt algorithm to Equation 4.25, in order to find the solution that minimizes the reprojection error. More precisely, we can find $\lambda$, $K$, $R$ and $\mathbf{t}$ that minimize

$$\sum_{i=1}^{n} d(\lambda K[R|t]X_i, x_i)^2.$$

## 4.9 API

```
void calib_singlecam( gsl_matrix *p, gsl_matrix *q,
                      gsl_matrix *x );
```

This function implements the algorithm of Section 4.1.2, $q$ receives the 3D points of the scene, and $x$ receives the respective projections. This representation puts each points and each projection in a line of the $q$ and $x$ matrices. The $3 \times 4$ camera is returned on the parameter $p$.

```
calib_get_normalize_transform( gsl_matrix *h, gsl_matrix
                               *hinv, gsl_matrix *s )
```

This function returns the homography and its inverse relative to the application of the algorithm described in the Section 4.2 to the 2D points of the rows of $s$.

```
void calib_apply_homog( gsl_matrix *r, gsl_matrix *h,
                        gsl_matrix *s )
```

This function returns in $r$ the application of the homography $h$ to the vector of the rows of the matrix $s$.

```
void calib_get_3dnormalize_transform( gsl_matrix *h,
                                      gsl_matrix *s )
```

This function returns the 3D normalization homography $h$ used to preprocessing the 3D points of the rows of $s$, such as explained in the Section 4.2.

```
void calib_apply_3dhomog( gsl_matrix *r, gsl_matrix *h,
                          gsl_matrix *s )
```

This function applies the homography $h$ to the points of the rows of $s$ and returns the new set of points in the rows of the matrix $r$

```
void calib_singlecam_dlt( gsl_matrix *p, gsl_matrix *q,
                          gsl_matrix *x );
```

Implements the algorithm of Section 4.1.1 with the same parameters of the function *calib_singlecam*. But, in this case, the 2D and 3D points are previously normalized such as explained in the Section 4.2.

```
void vector_cross( gsl_vector *r, gsl_vector *v1,
                   gsl_vector *v2 );
```

This function returns in $r$ the vector cross $v1 \times v2$.

```
void calib_camfactor( gsl_matrix *k, gsl_matrix *r,
                      gsl_vector *t, gsl_matrix *p );
```

This function implements the algorithm of Section 4.3. The camera is passed on the parameter $p$, and it returns the factorization of the camera $k[R|t]$ in the other 3 parameters.

```
void singlecam_nlin_calib( gsl_vector *xout,
gsl_vector *x, gsl_matrix *points, gsl_matrix *projs );
```

This function implements the algorithm of Section 4.5. The input parameters are encoded in the $x$ vector and the camera found is returned in $xout$. This is done, because the Levenberg-Marquardt algorithm of the GNU Scientific Library demands that the parameters are encoded on vectors. $x$ and $xout$ are encoded and decoded respecively by the functions *singlecam_set_krt* and *singlecam_get_krt*. To decode the $xout$ into a $3 \times 4$ matrix we can also use the function *singlecam_get_camera*.

```
int cost_func(const gsl_vector *x, void *params,
              gsl_vector *f);
```

This function implements the cost function used by the Levenberg-Marquardt algorithm in the function *singlecam_nlin_calib*.

```
double reproj_error( const gsl_vector *x,
gsl_matrix *points, gsl_matrix *projs, int point_index );
```

This function returns the error of the Levenberg-Marquardt algorithm relative to the camera encoded on the vector $x$, calculated considering the 3D points *points* and the projecitons *projs* passed as parameters considering the point index *point_index*. This function is used by the *cost_func*, inside an iteration that changes the *point_index* parameter considering all points. As a consequence the error vector of *cost_func* is filled.

```
gsl_vector* singlecam_param_alloc( void )
```

This function allocates and returns the parameters used in the non-linear calibration of a single camera. More precisely, it returns a vector with 10 elements used to represent the intrinsic and extrinsic of a camera used by a 3D graphical system.

```
void ba_axis_angle_to_r( gsl_matrix *r,
                         gsl_vector *axis_angle );
```

This function implements the algorithm of Section 4.5.2, converting the axis angle representation encoded in vector *axis_angle* to a $3 \times 3$ matrix $r$.

```
void ba_r_to_axis_angle( gsl_vector *axis_angle,
                         gsl_matrix *r );
```

This function implements the algorithm of Section 4.5.2, converting the $3 \times 3$ matrix $r$ to the axis angle representation *axis_angle*

```
void singlecam_get_krt( gsl_matrix *k, gsl_vector *r,
                        gsl_vector *t, const gsl_vector *x );
```

This function decodes the output of the Levenberg-Marquardt algorithm used for estimating one camera by the function *singlecam_nlin_calib*. the vector $x$ is decoded as a $k[R|t]$ camera in the other parameters.

```
void singlecam_set_krt( gsl_vector *x, gsl_matrix *k,
                        gsl_vector *r, gsl_vector *t );
```

This function encodes the initial value of the Levenberg-Marquardt algorithm executed by the function *singlecam_nlin_calib*. The vector $x$ is initialized with the camera $k[R|t]$ passed by the other parameters.

```
void singlecam_get_camera( gsl_matrix *camera,
                           const gsl_vector *x )
```

This function converts the output of the Levenberg-Marquardt algorithm executed by the function *singlecam_nlin_calib* into the form of a $3 \times 4$ camera matrix.

## 4.10 CODE

```
// calib/singlecam.c

void calib_singlecam( gsl_matrix *p, gsl_matrix *q, gsl_matrix *x )
{
 int n, i, j;
 double xi, yi, qXi, qYi, qZi;
 gsl_matrix *u, *v;
 gsl_vector *w, *s, *aux;
```

```
  n = x->size1;
  u = gsl_matrix_alloc( 2*n, 12 );
  v = gsl_matrix_alloc( 12, 12 );
  s = gsl_vector_alloc( 12 );
  aux = gsl_vector_alloc( 12 );
  w = gsl_vector_alloc( 12 );

  gsl_matrix_set_zero(u);
  for( i = 0; i < n; i++ ){
    xi = gsl_matrix_get( x, i, 0 );
    yi = gsl_matrix_get( x, i, 1 );
    qXi = gsl_matrix_get( q, i, 0 );
    qYi = gsl_matrix_get( q, i, 1 );
    qZi = gsl_matrix_get( q, i, 2 );
    gsl_matrix_set( u, 2*i, 0, qXi );
    gsl_matrix_set( u, 2*i, 1, qYi );
    gsl_matrix_set( u, 2*i, 2, qZi );
    gsl_matrix_set( u, 2*i, 3, 1 );
    gsl_matrix_set( u, 2*i + 1, 4, qXi );
    gsl_matrix_set( u, 2*i + 1, 5, qYi );
    gsl_matrix_set( u, 2*i + 1, 6, qZi );
    gsl_matrix_set( u, 2*i + 1, 7, 1 );
    gsl_matrix_set( u, 2*i, 8, -xi*qXi );
    gsl_matrix_set( u, 2*i, 9, -xi*qYi );
    gsl_matrix_set( u, 2*i, 10, -xi*qZi );
    gsl_matrix_set( u, 2*i, 11, -xi );
    gsl_matrix_set( u, 2*i + 1, 8, -yi*qXi );
    gsl_matrix_set( u, 2*i + 1, 9, -yi*qYi );
    gsl_matrix_set( u, 2*i + 1, 10, -yi*qZi );
    gsl_matrix_set( u, 2*i + 1, 11, -yi );
  }
  gsl_linalg_SV_decomp( u, v, s, w );
  gsl_matrix_get_col( aux, v, 11 );

  for( j=0; j < 3; j++ )
    for( i=0; i < 4; i++ )
      gsl_matrix_set( p, j, i, gsl_vector_get( aux, 4*j + i ) );

  gsl_matrix_free(u);
  gsl_matrix_free(v);
  gsl_vector_free(s);
  gsl_vector_free(aux);
  gsl_vector_free(w);
}


// calib/normalize.c

void calib_get_normalize_transform( gsl_matrix *h, gsl_matrix *hinv,
                                    gsl_matrix *s )
{
  int i, n;
  double d = 0, x0 = 0, y0 = 0, scale;

  n = s->size1;
  for( i = 0; i < n; i++ ){
    x0 += gsl_matrix_get( s, i, 0);
    y0 += gsl_matrix_get( s, i, 1);
  }
  x0 = x0/n;
  y0 = y0/n;

  for( i = 0; i < n; i++ ){
    d += SQR(gsl_matrix_get(s, i, 0) - x0) +
```

```
        SQR(gsl_matrix_get(s, i, 1) - y0);
 }

 gsl_matrix_set_zero(h);
 scale = sqrt( 2*n/d );
 gsl_matrix_set( h, 0, 0, scale );
 gsl_matrix_set( h, 1, 1, scale );
 gsl_matrix_set( h, 0, 2, -scale*x0 );
 gsl_matrix_set( h, 1, 2, -scale*y0 );
 gsl_matrix_set( h, 2, 2, 1. );

 if( hinv != NULL ){
   gsl_matrix_set_zero(hinv);
   gsl_matrix_set( hinv, 0, 0, 1/scale );
   gsl_matrix_set( hinv, 1, 1, 1/scale );
   gsl_matrix_set( hinv, 0, 2, x0 );
   gsl_matrix_set( hinv, 1, 2, y0 );
   gsl_matrix_set( hinv, 2, 2, 1. );
 }
}
```

```
// calib/homog.c

void calib_apply_homog( gsl_matrix *r, gsl_matrix *h, gsl_matrix *s )
{
 int i;
 gsl_matrix *v1, *v2;

 v1 = gsl_matrix_alloc( 3, 1 );
 v2 = gsl_matrix_alloc( 3, 1 );
 for( i = 0; i < s->size1; i++ ){
   gsl_matrix_set( v1, 0, 0, gsl_matrix_get( s, i, 0 ) );
   gsl_matrix_set( v1, 1, 0, gsl_matrix_get( s, i, 1 ) );
   gsl_matrix_set( v1, 2, 0, 1. );
   gsl_linalg_matmult( h, v1, v2 );
   gsl_matrix_set( r, i, 0, gsl_matrix_get(v2, 0, 0)/gsl_matrix_get(v2, 2, 0));
   gsl_matrix_set( r, i, 1, gsl_matrix_get(v2, 1, 0)/gsl_matrix_get(v2, 2, 0));
 }

 gsl_matrix_free(v1);
 gsl_matrix_free(v2);
}
```

```
// calib/normalize3d.c

void calib_get_3dnormalize_transform( gsl_matrix *h, gsl_matrix *s )
{
 int i, n;
 double d = 0, x0 = 0, y0 = 0, z0 = 0, scale;

 n = s->size1;
 for( i = 0; i < n; i++ ){
   x0 += gsl_matrix_get( s, i, 0);
   y0 += gsl_matrix_get( s, i, 1);
   z0 += gsl_matrix_get( s, i, 2);
 }
 x0 = x0/n;
 y0 = y0/n;
 z0 = z0/n;

 for( i = 0; i < n; i++ ){
   d += SQR(gsl_matrix_get(s, i, 0) - x0) +
```

```
        SQR(gsl_matrix_get(s, i, 1) - y0) +
        SQR(gsl_matrix_get(s, i, 2) - z0);
 }

 gsl_matrix_set_zero(h);
 scale = sqrt( 3*n/d );
 gsl_matrix_set( h, 0, 0, scale );
 gsl_matrix_set( h, 1, 1, scale );
 gsl_matrix_set( h, 2, 2, scale );
 gsl_matrix_set( h, 0, 3, -scale*x0 );
 gsl_matrix_set( h, 1, 3, -scale*y0 );
 gsl_matrix_set( h, 2, 3, -scale*z0 );
 gsl_matrix_set( h, 3, 3, 1. );
}


// calib/homog3d.c

void calib_apply_3dhomog( gsl_matrix *r, gsl_matrix *h, gsl_matrix *s )
{
 int i;
 gsl_matrix *v1, *v2;

 v1 = gsl_matrix_alloc( 4, 1 );
 v2 = gsl_matrix_alloc( 4, 1 );
 for( i = 0; i < s->size1; i++ ){
   gsl_matrix_set( v1, 0, 0, gsl_matrix_get( s, i, 0 ) );
   gsl_matrix_set( v1, 1, 0, gsl_matrix_get( s, i, 1 ) );
   gsl_matrix_set( v1, 2, 0, gsl_matrix_get( s, i, 2 ) );
   gsl_matrix_set( v1, 3, 0, 1. );
   gsl_linalg_matmult( h, v1, v2 );
   gsl_matrix_set( r, i, 0, gsl_matrix_get(v2, 0, 0)/gsl_matrix_get(v2, 3, 0));
   gsl_matrix_set( r, i, 1, gsl_matrix_get(v2, 1, 0)/gsl_matrix_get(v2, 3, 0));
   gsl_matrix_set( r, i, 2, gsl_matrix_get(v2, 2, 0)/gsl_matrix_get(v2, 3, 0));
 }

 gsl_matrix_free(v1);
 gsl_matrix_free(v2);
}


void calib_singlecam_dlt( gsl_matrix *p, gsl_matrix *q, gsl_matrix *x )
{
 gsl_matrix *h, *hinv, *homog3d, *paux, *paux_homog3d, *qn, *xn;

 h = gsl_matrix_alloc( 3, 3 );
 hinv = gsl_matrix_alloc( 3, 3 );
 homog3d = gsl_matrix_alloc( 4, 4 );
 paux = gsl_matrix_alloc( 3, 4 );
 paux_homog3d = gsl_matrix_alloc( 3, 4 );
 qn = gsl_matrix_alloc( q->size1, 3 );
 xn = gsl_matrix_alloc( x->size1, 2 );

 calib_get_normalize_transform( h, hinv, x );
 calib_get_3dnormalize_transform( homog3d, q );
 calib_apply_homog( xn, h, x );
 calib_apply_3dhomog( qn, homog3d, q );
 calib_singlecam( paux, qn, xn );
 gsl_linalg_matmult( paux, homog3d, paux_homog3d );
 gsl_linalg_matmult( hinv, paux_homog3d, p );

 gsl_matrix_free( h );
 gsl_matrix_free( hinv );
 gsl_matrix_free( homog3d );
```

```
 gsl_matrix_free( paux );
 gsl_matrix_free( paux_homog3d );
 gsl_matrix_free( qn );
 gsl_matrix_free( xn );
}

//calib/camfactor.c

void vector_cross( gsl_vector *r, gsl_vector *v1, gsl_vector *v2 )
{
 double v10, v11, v12,
        v20, v21, v22;

 v10 = gsl_vector_get( v1, 0 );
 v11 = gsl_vector_get( v1, 1 );
 v12 = gsl_vector_get( v1, 2 );
 v20 = gsl_vector_get( v2, 0 );
 v21 = gsl_vector_get( v2, 1 );
 v22 = gsl_vector_get( v2, 2 );

 gsl_vector_set( r, 0, v11*v22 - v12*v21 );
 gsl_vector_set( r, 1, v12*v20 - v10*v22 );
 gsl_vector_set( r, 2, v10*v21 - v11*v20 );
}


void calib_camfactor( gsl_matrix *k, gsl_matrix *r,
                      gsl_vector *t, gsl_matrix *p )
{
 int i;
 double lambda, v0, u0, s, f1, f2, t1, t2, t3;
 gsl_matrix *p_til;
 gsl_vector *a1, *a2, *a3, *aux, *r1;

 p_til = gsl_matrix_alloc( 3, 4 );
 a1 = gsl_vector_alloc(3);
 a2 = gsl_vector_alloc(3);
 a3 = gsl_vector_alloc(3);
 aux = gsl_vector_alloc(3);
 r1 = gsl_vector_alloc(3);

 lambda = sqrt( SQR(gsl_matrix_get( p, 2, 0 )) +
                SQR(gsl_matrix_get( p, 2, 1 )) +
                SQR(gsl_matrix_get( p, 2, 2 )) );

 gsl_matrix_memcpy( p_til, p );
 gsl_matrix_scale( p_til, 1/lambda );

 for( i = 0; i < 3; i++ ){
   gsl_vector_set( a1, i, gsl_matrix_get( p_til, 0, i ) );
   gsl_vector_set( a2, i, gsl_matrix_get( p_til, 1, i ) );
   gsl_vector_set( a3, i, gsl_matrix_get( p_til, 2, i ) );
 }

 gsl_matrix_set_row( r, 2, a3 );
 gsl_blas_ddot( a2, a3, &v0 );
 gsl_blas_ddot( a1, a3, &u0 );

 gsl_vector_memcpy( aux, a3 );
 gsl_vector_scale( aux, -v0 );
 gsl_vector_add( aux, a2 );
 f2 = gsl_blas_dnrm2( aux );
 gsl_vector_scale( aux, 1/f2 );
 gsl_matrix_set_row( r, 1, aux );
```

```
 gsl_blas_ddot( a1, aux, &s );
 vector_cross( r1 ,aux, a3 );
 gsl_matrix_set_row( r, 0, r1 );
 gsl_blas_ddot( a1, r1, &f1 );

 t3 = gsl_matrix_get( p_til, 2, 3 );
 t2 = (gsl_matrix_get( p_til, 1, 3 ) - v0*t3)/f2;
 t1 = (gsl_matrix_get( p_til, 0, 3) - u0*t3 - s*t2 )/f1;

 if( f1 < 0 ){
   t2 = -t2;
   t3 = -t3;
   f1 = -f1;
   gsl_vector_scale( a3, -1 );
   gsl_matrix_set_row( r, 2, a3 );
   gsl_matrix_get_row( aux, r, 1 );
   gsl_vector_scale( aux, -1 );
   gsl_matrix_set_row( r, 1, aux );
 }

 gsl_vector_set( t, 0, t1 );
 gsl_vector_set( t, 1, t2 );
 gsl_vector_set( t, 2, t3 );

 gsl_matrix_set_identity(k);
 gsl_matrix_set( k, 0, 0, f1 );
 gsl_matrix_set( k, 1, 1, f2 );
 gsl_matrix_set( k, 0, 1, s );
 gsl_matrix_set( k, 0, 2, u0 );
 gsl_matrix_set( k, 1, 2, v0 );

 gsl_matrix_free( p_til );
 gsl_vector_free( a1 );
 gsl_vector_free( a2 );
 gsl_vector_free( a3 );
 gsl_vector_free( aux );
 gsl_vector_free( r1 );
}

// singlecam/singlecam.c

void singlecam_nlin_calib( gsl_vector *xout, gsl_vector *x,
                           gsl_matrix *points, gsl_matrix *projs )
{
 const gsl_multifit_fdfsolver_type *t = gsl_multifit_fdfsolver_lmsder;
 gsl_multifit_fdfsolver *s;
 gsl_multifit_function_fdf f;
 int npoints = projs->size1;
 SingleCamData d;

 d.projs = projs;
 d.points = points;

 f.f = &cost_func;
 f.df = NULL;

 f.p = x->size;
 f.n = npoints;

 f.params = &d;

 s = gsl_multifit_fdfsolver_alloc(t, f.n, f.p);
 gsl_multifit_fdfsolver_set(s, &f, x);
```

```
 ba_optimize(s);

 gsl_vector_memcpy( xout, s->x );
 gsl_multifit_fdfsolver_free(s);
}


int cost_func(const gsl_vector *x, void *params, gsl_vector *f)
{
 int j;
 gsl_matrix *projs;
 gsl_matrix *points;

 projs = ((SingleCamData*)params)->projs;
 points = ((SingleCamData*)params)->points;
 gsl_vector_set_zero(f);
 for( j = 0; j < projs->size1; j++ )
    gsl_vector_set( f, j, reproj_error( x, points, projs, j ) );

 return GSL_SUCCESS;
}


double reproj_error( const gsl_vector *x, gsl_matrix *points,
                     gsl_matrix *projs, int point_index )
{
 double error;
 gsl_vector  *prj, *v, *v_prj;
 gsl_matrix *p;

 v = gsl_vector_alloc(3);
 v_prj = gsl_vector_alloc(2);
 prj = gsl_vector_alloc(2);
 p = gsl_matrix_alloc( 3,4 );

 gsl_matrix_get_row( prj, projs, point_index );
 gsl_matrix_get_row( v, points, point_index );
 singlecam_get_camera( p, x );
 calib_apply_P( p, v, v_prj );
 gsl_vector_sub( prj, v_prj );
 error = SQR( gsl_blas_dnrm2( prj ) );

 gsl_vector_free(v);
 gsl_vector_free(v_prj);
 gsl_vector_free(prj);
 gsl_matrix_free(p);

 return error;
}


// singlecam/coder.c

#include "singlecam.h"

static double kvalues[5];

/* Define the free parameters used by the optimization process */
/* It forces the intrinsic parameter to be compatible with the S3D Library */
static int free_kparams[] = {1,0,1,1,1};


gsl_vector* singlecam_param_alloc( void )
{
```

```
    int i, n = 0;

    for( i=0; i<5; i++ )
      if( free_kparams[i] != 0 )
        n++;

    return  gsl_vector_alloc( 6 + n );
}


// ba/rotations.c

void ba_axis_angle_to_r( gsl_matrix *r, gsl_vector *axis_angle )
{
 double a1, a2, a3, s, c, angle;
 gsl_vector *a;

 a = gsl_vector_alloc(3);

 angle =  gsl_blas_dnrm2( axis_angle );

 if( angle <= ANG_EPS )
    gsl_matrix_set_identity(r);
 else{
   s = sin(angle);
   c = cos(angle);
   gsl_vector_memcpy( a, axis_angle );
   gsl_vector_scale( a, 1./angle );

   a1 = gsl_vector_get( a, 0 );
   a2 = gsl_vector_get( a, 1 );
   a3 = gsl_vector_get( a, 2 );

   gsl_matrix_set( r, 0, 0, a1*a1 + c*( 1 - a1*a1 ) );
   gsl_matrix_set( r, 1, 0, a1*a2*( 1 - c ) + a3*s );
   gsl_matrix_set( r, 2, 0, a1*a3*( 1 - c ) - a2*s );

   gsl_matrix_set( r, 0, 1, a1*a2*( 1 - c ) - a3*s );
   gsl_matrix_set( r, 1, 1, a2*a2 + c*( 1 - a2*a2 ) );
   gsl_matrix_set( r, 2, 1, a2*a3*( 1 - c ) + a1*s );

   gsl_matrix_set( r, 0, 2, a1*a3*( 1 - c ) + a2*s );
   gsl_matrix_set( r, 1, 2, a2*a3*( 1 - c ) - a1*s );
   gsl_matrix_set( r, 2, 2, a3*a3 + c*( 1 - a3*a3 ) );
 }

 gsl_vector_free(a);
}

void ba_r_to_axis_angle( gsl_vector *axis_angle, gsl_matrix *r )
{
 double s_doub, c_doub, angle;
 gsl_vector *w, *s;
 gsl_matrix *m, *v;

 m = gsl_matrix_alloc(3,3);
 v = gsl_matrix_alloc(3,3);
 s = gsl_vector_alloc(3);
 w = gsl_vector_alloc(3);

 gsl_matrix_set_identity( m );
 gsl_matrix_scale( m, -1. );
 gsl_matrix_add( m, r );
```

```
 gsl_linalg_SV_decomp( m, v, s,w );
 gsl_matrix_get_col( axis_angle, v, 2 );
 gsl_vector_scale( axis_angle, 1/gsl_blas_dnrm2( axis_angle ) );

 s_doub = gsl_vector_get( axis_angle, 0 )*( gsl_matrix_get( r, 2,1 )
                         - gsl_matrix_get( r, 1, 2 )) +
          gsl_vector_get( axis_angle, 1 )*( gsl_matrix_get( r, 0,2 )
                         - gsl_matrix_get( r, 2, 0 )) +
          gsl_vector_get( axis_angle, 2 )*( gsl_matrix_get( r, 1,0 )
                         - gsl_matrix_get( r, 0, 1 ));

 c_doub = gsl_matrix_get( r, 0,0 ) + gsl_matrix_get( r, 1, 1 )
                         + gsl_matrix_get( r, 2, 2 ) - 1;
 angle = atan2( s_doub, c_doub );

 gsl_vector_scale( axis_angle, angle );

 gsl_matrix_free(m);
 gsl_matrix_free(v);
 gsl_vector_free(s);
 gsl_vector_free(w);
}


void singlecam_get_krt( gsl_matrix *k, gsl_vector *r,
                        gsl_vector *t, const gsl_vector *x )
{
 int i, j, p = 0, q = 0;

 gsl_matrix_set_identity( k );

 gsl_vector_set( r, 0, gsl_vector_get( x, 0 ) );
 gsl_vector_set( r, 1, gsl_vector_get( x, 1 ) );
 gsl_vector_set( r, 2, gsl_vector_get( x, 2 ) );
 gsl_vector_set( t, 0, gsl_vector_get( x, 3 ) );
 gsl_vector_set( t, 1, gsl_vector_get( x, 4 ) );
 gsl_vector_set( t, 2, gsl_vector_get( x, 5 ) );

 for( i=0; i < 2; i++ )
   for( j = i; j < 3; j++ ){
     if( free_kparams[p] == 1 ){
       gsl_matrix_set( k, i, j, gsl_vector_get( x, p+6 ) );
       p++;
     }
     else
       gsl_matrix_set( k, i, j, kvalues[q++] );
   }
}


void singlecam_set_krt( gsl_vector *x, gsl_matrix *k,
                        gsl_vector *r, gsl_vector *t )
{
 int i, j, p = 0, q = 0;

 gsl_vector_set( x, 0, gsl_vector_get( r, 0 ) );
 gsl_vector_set( x, 1, gsl_vector_get( r, 1 ) );
 gsl_vector_set( x, 2, gsl_vector_get( r, 2 ) );
 gsl_vector_set( x, 3, gsl_vector_get( t, 0 ) );
 gsl_vector_set( x, 4, gsl_vector_get( t, 1 ) );
 gsl_vector_set( x, 5, gsl_vector_get( t, 2 ) );


 for( i=0; i < 2; i++ )
```

```
    for( j = i; j < 3; j++ ){
      if( free_kparams[p] == 1 ){
        gsl_vector_set( x, p+6, gsl_matrix_get( k, i, j ) );
        p++;
      }
      else
        kvalues[q++] = gsl_matrix_get( k, i, j );
    }
}


void singlecam_get_camera( gsl_matrix *camera, const gsl_vector *x )
{
 gsl_vector *h, *t;
 gsl_matrix *k, *r;

 h = gsl_vector_alloc(3);
 t = gsl_vector_alloc(3);
 k = gsl_matrix_alloc(3,3);
 r = gsl_matrix_alloc(3,3);

 singlecam_get_krt( k, h, t, x );
 ba_axis_angle_to_r( r, h );
 calib_pmatrix_make( camera, k, r, t );

 gsl_vector_free(h);
 gsl_vector_free(t);
 gsl_matrix_free(k);
 gsl_matrix_free(r);
}
```

## 4.11 SINGLE CAMERA CALIBRATION PROGRAM

This program returns the intrinsic and extrinsic parameters given the correspondence of a set of 3D points with those projections made by the camera. The result is calculated using the Levenberg-Marquardt, forcing the intrinsic parameters to be compatible with the ones used by the S3D Library.

argv[1] is the name of a file that contain the projection points coordinates, and argv[2] is the name of a file that contains the 3D points coordinates. The camera $k[R|t]$ correspondent to the result is encoded in three files with the names: $k$, $r$ and $t$.

```
// singlecam_nlin/main.c

#include "singlecam.h"

static int get_nlines( FILE *f );

int main( int argc, char **argv )
{
 int npoints;
 FILE *fprojs, *fpoints;
 FILE *fk, *fr, *ft;
 gsl_matrix *projs, *points, *p;
 gsl_vector *x, *xout, *r_vec, *t;
 gsl_matrix *k, *r;

 k = gsl_matrix_alloc( 3, 3 );
```

```
 r = gsl_matrix_alloc( 3, 3 );
 r_vec = gsl_vector_alloc( 3 );
 t = gsl_vector_alloc( 3 );
 p = gsl_matrix_alloc( 3, 4 );
 fprojs = fopen( argv[1], "r" );
 npoints = get_nlines( fprojs );
 fclose( fprojs );
 projs = gsl_matrix_alloc( npoints, 2 );
 points = gsl_matrix_alloc( npoints, 3 );
 x = singlecam_param_alloc();
 xout =  singlecam_param_alloc();

 fprojs = fopen( argv[1], "r" );
 fpoints = fopen( argv[2], "r" );
 gsl_matrix_fscanf( fprojs, projs );
 gsl_matrix_fscanf( fpoints, points );
 fclose( fprojs );
 fclose( fpoints );

 calib_singlecam_dlt( p, points, projs );
 calib_camfactor( k, r, t, p );
 gsl_matrix_set( k, 0, 1, 0 );
 ba_r_to_axis_angle( r_vec, r );
 singlecam_set_krt( x, k, r_vec, t );
 singlecam_nlin_calib( xout, x, points, projs );
 singlecam_get_krt( k, r_vec, t, xout );
 ba_axis_angle_to_r( r, r_vec );

 fk = fopen( "k", "w" );
 fr = fopen( "r", "w" );
 ft = fopen( "t", "w" );
 gsl_matrix_fprintf( fk, k, "%f" );
 gsl_matrix_fprintf( fr, r, "%f" );
 gsl_vector_fprintf( ft, t, "%f" );
 fclose(fk);
 fclose(fr);
 fclose(ft);

 gsl_matrix_free(p);
 gsl_matrix_free(projs);
 gsl_matrix_free(points);
 gsl_vector_free(x);
 gsl_vector_free(xout);
 gsl_matrix_free(k);
 gsl_matrix_free(r);
 gsl_vector_free(r_vec);
 gsl_vector_free(t);
}


int get_nlines( FILE *f )
{
 char c;
 int nlines = 0;
 double x1, y1;

 while( !feof( f ) ){
   if( (c = getc(f)) != '%' ){
     ungetc( c, f );
     fscanf( f, "%lf %lf\n", &x1, &y1);
     nlines++;
   }
 }
 return nlines;
```

## 4.12 SIX POINTS SINGLE CAMERA CALIBRATION PROGRAM

Sometimes we prefer to estimate a camera without computing the non linear optimization step, instead we can calculate it just using the solution presented in the Sections 4.1.1 and 4.3. Although this solution can be less accurate, it demands just 6 correspondences between 3d points and image points, and its precision gives a result that can be good enough for some applications. It happens, for example, in the case of Image-Based Lighting as we will see in the Chapter 10. The code of this algorithm is presented below:

```
// singlecam/main.c

#include "singlecam.h"

static int get_nlines( FILE *f );

int main( int argc, char **argv )
{
 int npoints;
 FILE *fprojs, *fpoints;
 FILE *fk, *fr, *ft;
 gsl_matrix *projs, *points, *p;
 gsl_vector *x, *xout, *r_vec, *t;
 gsl_matrix *k, *r;

 k = gsl_matrix_alloc( 3, 3 );
 r = gsl_matrix_alloc( 3, 3 );
 r_vec = gsl_vector_alloc( 3 );
 t = gsl_vector_alloc( 3 );
 p = gsl_matrix_alloc( 3, 4 );
 fprojs = fopen( argv[1], "r" );
 npoints = get_nlines( fprojs );
 fclose( fprojs );
 projs = gsl_matrix_alloc( npoints, 2 );
 points = gsl_matrix_alloc( npoints, 3 );
 x = singlecam_param_alloc();
 xout =  singlecam_param_alloc();

 fprojs = fopen( argv[1], "r" );
 fpoints = fopen( argv[2], "r" );
 gsl_matrix_fscanf( fprojs, projs );
 gsl_matrix_fscanf( fpoints, points );
 fclose( fprojs );
 fclose( fpoints );

 calib_singlecam_dlt( p, points, projs );
 calib_camfactor( k, r, t, p );

 fk = fopen( "k", "w" );
 fr = fopen( "r", "w" );
 ft = fopen( "t", "w" );
 gsl_matrix_fprintf( fk, k, "%f" );
 gsl_matrix_fprintf( fr, r, "%f" );
 gsl_vector_fprintf( ft, t, "%f" );
 fclose(fk);
 fclose(fr);
 fclose(ft);

 gsl_matrix_free(p);
 gsl_matrix_free(projs);
```

```
 gsl_matrix_free(points);
 gsl_vector_free(x);
 gsl_vector_free(xout);
 gsl_matrix_free(k);
 gsl_matrix_free(r);
 gsl_vector_free(r_vec);
 gsl_vector_free(t);
}


int get_nlines( FILE *f )
{
 char c;
 int nlines = 0;
 double x1, y1;

 while( !feof( f ) ){
   if( (c = getc(f)) != '%' ){
     ungetc( c, f );
     fscanf( f, "%lf %lf\n", &x1, &y1);
     nlines++;
   }
 }
 return nlines;
}
```

CHAPTER 5

# Estimating Two Cameras

THE PURPOSE of this chapter is to describe an algorithm that determines the position relative between the cameras that were used to capture two images. Most precisely, we are interested in solving the following problem:

**Problem 5.0.1.** *Given a set* $(x_1, \hat{x}_1), (x_{21}, \hat{x}_2), ..., (x_n, \hat{x}_n)$ *with* $(x_i, \hat{x}_i) \in \mathbb{R}P^2$, *which corresponds to the projections in a pair of images* $I_1$ *and* $I_2$, *of a set of scene points* $X_1, X_2, ..., X_n$, *with* $X_i \in \mathbb{R}P^3$. *Determine relative placement between the cameras that captured* $I_1$ *and* $I_2$, *assuming that the intrinsic parameters of both are known.*

The elements of the pair $(x_i, \hat{x}_i)$ are called homologous points associated with $X_i$.

## 5.1 REPRESENTATION OF RELATIVE POSITIONING

In order to represent the relative positioning between two cameras we will assume, without loss of generality, that one of the cameras is $K_1[I|0]$ which corresponds to a camera positioned at the origin with a view direction in the direction of the z-axis. In this way the extrinsic parameters of the other camera, $K_2[R|t]$, specify the relative positioning between them.

An important fact is that the projections of a set of points in the scene $\{X_1, X_2, ..., X_n\}$, with $X_i \in \mathbb{R}^3$ , relative to the cameras $K_1[I|0]$ and $K_2[R|t]$ are equal to the projections of the set $\{\lambda X_1, \lambda X_2, \ldots, \lambda X_n\}$ relative to the cameras $K_1[I|0]$ and $K_2[R|\lambda t]$, with $\lambda \in \mathbb{R}^+$ . In fact, it is enough to observe the following equalities:

$$K_1[I|0](\lambda X_i^T, 1)^T = K_1(\lambda X_i) = K_1(X_i) = K_1[I|0](X_i^T, 1)^T,$$

$$\begin{aligned} K_2[R|\lambda t](\lambda X_i^T, 1)^T &= K_2(R(\lambda X_i) + \lambda t) = K_2\lambda(RX_i + t) \\ &= K_2(RX_i + t) = K_2[R|t](X_i^T, 1)^T. \end{aligned}$$

This shows that the Problem 5.0.1 is defined with an ambiguity of scale, because the value of $||t||$ cannot be determined.

DOI: 10.1201/9781003206026-5

## 5.2 RIGID MOVEMENT

The theorem below, presented in [23], establishes a constraint for defined in two $\mathbb{R}^3$ references, which are related by a rigid movement.

**Theorem 5.1.** *Let $X, \hat{X} \in \mathbb{R}^3$ defined so that $\hat{X} = RX + t$, where $R$ is the rotation matrix, and $t \in \mathbb{R}^3$. If $[t]_x : \mathbb{R}^3 \to \mathbb{R}^3$ is the linear operator defined by $[t]_x(x) = t \times x$, then $\hat{X}^T([t]_x R)X = 0$.*

**Proof**

Using the fact that the vector $\hat{X} \times t$ is perpendicular to both $\hat{X}$ and $t$, we have that

$$(\hat{X} \times t) \cdot \hat{X} = 0 \text{ and } (\hat{X} \times t) \cdot t = 0.$$

As a consequence:

$$\hat{X}^T([t]_x R)X = \hat{X} \cdot (t \times RX) = (\hat{X} \times t) \cdot (RX) = (\hat{X} \times t) \cdot (RX + t) = (\hat{X} \times t) \cdot \hat{X} = 0.$$

## 5.3 OTHER PROJECTIVE MODEL

The effect obtained by the projective camera, defined by the transformation $[I|0]$, is equivalent to the effect of the transformation $T : \mathbb{R}^3 \to \mathbb{R}P^2$, which applies each point $x \in \mathbb{R}^3$ in a point of $\mathbb{R}P^2$, whose homogeneous coordinates are $\lambda x$, such that $\lambda \in \mathbb{R} - \{0\}$. In both cases the effect is the same as that of the projection projection $T : \mathbb{R}^3 \to \mathbb{R}^2$ defined by

$$(x, y, z) \mapsto (\frac{x}{z}, \frac{y}{z}).$$

## 5.4 EPIPOLAR GEOMETRY

Epipolar Geometry is the study of the geometric relationships existing between the projections of a set of points on two images obtained by projective cameras.

Next, an algebraic development of Epipolar Geometry will be done. Initially will be considered the case in which the cameras are of the form $[I|0]$ and $[R|t]$, that is, the matrix of the intrinsic parameters of both cameras is the identity matrix. In this case the relationships of epipolarity will be characterized by the Essential Matrix.

Subsequently, the general case will be dealt with, in which the cameras are of the form $K_1[I|0]$ and $K_2[R|t]$. In this case, the epipolarity relations will be characterized by the Fundamental Matrix.

### 5.4.1 Essential Matrix

Defining $E = [t]_\times R$, we have from Theorem 5.1 that the expression $\hat{X}^T EX = 0$, which relates the coordinates, in $\mathbb{R}^3$, of a point in the scene in the referentials associated with $[I|0]$ and $[R|t]$ cameras. To obtain a relationship between the coordinates of the projections of that point in the images captured by these cameras just note that for all $\lambda_1, \lambda_2 \in R - \{0\}$

$$\hat{X}^T E X = 0 \Leftrightarrow (\lambda_1 \hat{X}^T) E (\lambda_2 X) = 0.$$

Since $X$ and $\hat{X}$ are 3D coordinates defined in the referentials of both cameras, we can use the result presented on Section 5.3 to conclude that, if $x \in \mathbb{R}P^2$ and $\hat{x} \in \mathbb{R}P^2$ are the homogeneous coordinates of the projections of a point in the scene obtained by the cameras $[I|0]$ and $[R|t]$ respectively, then the relation $\hat{x}^T E x = 0$ is valid. It is also easy to see that the matrix $E$, called an essential matrix, is defined in less than one product by a number.

## 5.5 FUNDAMENTAL MATRIX

Let us now consider that $x \in \mathbb{R}P^2$ is the projection of a point $X \in \mathbb{R}P^3$ obtained by the camera $K[R|t]$. The projection of the same point $X$ obtained by the camera $[R|t]$ is given by $K^{-1}x$. With this result we can generalize the relationship established by the essential matrix for the case where the cameras do not have the intrinsic parameters matrix equal to the identity. More precisely, given two cameras $K_1[I|0]$ and $K_2[R|t]$, we have that projections of a point $X$ relative to these cameras are $x$ and $\hat{x}$, respectively, so this relationship is valid

$$(K_2^{-1}\hat{x})^T([t]_\times R)(K_1^{-1}x) = 0. \tag{5.1}$$

This relationship can be rewritten as

$$\hat{x} F x = 0, \tag{5.2}$$

such that

$$F = K_2^{-T}[t]_\times R K^{-1} \tag{5.3}$$

is a $3 \times 3$ matrix, called the fundamental matrix.

## 5.6 THE 8-POINTS ALGORITHM

The 8-points algorithm was initially introduced in [18]. Your input is a set of pairs of homologous points $\{(x_1, \hat{x}_1), (x_2, \hat{x}_2), ..., (x_n, \hat{x}_n)\}$ defined over two images, and your answer is the fundamental matrix associated with the image pair. Your name is due to the fact that at least 8 pairs of homologous points are needed so that the algorithm can be executed. It consists of two stages:

1. Step 1: Determination of matrix $F$, which best satisfies $\hat{x}_i^T F x_i = 0$, for all $i \in \{1, 2, \ldots, n\}$.

2. Step 2: Determination of the matrix $\tilde{F}$ that is closest to $F$, and that satisfies $det(\tilde{F}) = 0$. The matrix $\tilde{F}$ is the output of the algorithm.

The details of the execution of the two steps, as well as the precise meaning of the expressions "best satisfies" and "closest," will be presented next.

### 5.6.1 Calculation of F

Considering that each of the elements of $F$ is a variable, and that the values of $x_i$ and $\hat{x}_i$ are known for each $i \in \{1, 2, 3, \ldots, n\}$, it follows that the expression $\hat{x}_i^T F x_i = 0$ defines a linear equation over 9 variables.

If $F_0$ is a solution for the previous equation, then $\lambda F_0$ is also a solution for any $\lambda \in \mathbb{R} - \{0\}$. This shows that it is sufficient to use a set of 8 pairs of points to form a linear system that allows the determination of the $F$ value. By similar reasons to those presented in the estimation of cameras, in Section 4.1.1, we have that the solution obtained using only 8 pairs of homologous points is not good, being interesting to use a larger set of points, converting the problem into an optimization problem.

### 5.6.2 Using More Than 8 Points

In order to solve the linear system defined by the equations $\hat{x}_i^T F x_i = 0$, using more than 8 pairs of homologous points, the problem can be reformulated as being that of find the $F_0$ matrix that minimizes the objective function

$$F \mapsto \sum_{i=1}^{n} \hat{x}_i^T F x_i,$$

that can be solved by Theorem 3.2. It is enough to rewritten it in the form $\min_{\|F\|=1} \|AF\|$, such that $F = (F_{11}, F_{12}, F_{13}, F_{21}, F_{22}, F_{23}, F_{31}, F_{32}, F_{33})^T$ and $A$ defined by

$$\begin{pmatrix} u'_1u_1 & v'_1u_1 & u_1 & u'_1v_1 & v'_1v_1 & v_1 & u'_1 & v'_1 & 1 \\ u'_2u_2 & v'_2u_2 & u_2 & u'_2v_2 & v'_2v_2 & v_2 & u'_2 & v'_2 & 1 \\ u'_3u_3 & v'_3u_3 & u_3 & u'_3v_3 & v'_3v_3 & v_3 & u'_3 & v'_3 & 1 \\ \vdots & \vdots & \vdots & \vdots & \vdots & \vdots & \vdots & \vdots & \vdots \\ u'_mu_m & v'_mu_m & u_m & u'_mv_m & v'_mv_m & v_m & u'_m & v'_m & 1 \end{pmatrix},$$

such that $x_i = (u_i, v_i, 1)^T$ and $x_i = (u'_i, v'_i, 1)^T$.

The constraint $\|F\| = 1$ makes sense because the fundamental matrices are defined at less than a multiplication by a scalar.

### 5.6.3 Calculation of $\tilde{F}$

The purpose of the $\tilde{F}$ calculation is to enforce the response of the 8-point algorithm satisfies an important property of the fundamental matrices, which is the fact that they are singular matrices [13]. This constraint is not imposed during the calculation of $F$.

We can define $\tilde{F}$ as being the singular matrix such that $\|F - \tilde{F}\|$ assumes the minimum value. Considering the norm used as being the Frobenius norm, there is a simple way to calculate $\tilde{F}$, which consists of directly using the proposition below, whose demonstration can be found in [29].

**Theorem 5.2.** *If $U diag(r, s, t)V^T$ is the SVD decomposition of $F$, with $r \geqslant s \geqslant t$, then the singular matrix $\tilde{F}$, such that $\|\tilde{F} - F\|$ is minimal, is given by $\tilde{F} = U diag(r, s, 0)V^T$.*

## 5.7 NORMALIZED 8-POINTS ALGORITHM

The 8-points algorithm is poorly conditioned. A simple modification that makes it better conditioned is described in [14]. The modification consists of applying two transformations $H_1 : \mathbb{R}P^2 \to \mathbb{R}P^2$ and $H_2 : \mathbb{R}P^2 \to \mathbb{R}P^2$ to the homologous points of the input set $A = \{(x_1, \hat{x}_1), (x_2, \hat{x}_2), \ldots, (x_n, \hat{x}_n)\}$, transforming it into the set $B = \{(H_1 x_1, H_2 \hat{x}_1), (H_1 x_2, H_2 \hat{x}_2), \ldots, (H_1 x_n, H_2 \hat{x}_n)\}$, where $H_1$ and $H_2$ are defined in to satisfy the following properties:

1. $H_1$ and $H_2$ are affine transformations that perform translation and scaling in $\mathbb{R}^2$.
2. Both sets, $H_1 x_1, H_1 x_2, \ldots, H_1 x_n$ and $H_2 \hat{x}_1, H_2 \hat{x}_2, \ldots, H_2 \hat{x}_n$, have the point $(0, 0)^T \in \mathbb{R}^2$ as the centroid.
3. The RMS value of the distances from the points of both sets, $H_1 x_1, H_1 x_2, \ldots, H_1$ $x_n$ and $H_2 \hat{x}_1, H_2 \hat{x}_2, \ldots, H_2 \hat{x}_n$, to the point $(0, 0)^T$ is $\sqrt{2}$.

The 8-point algorithm estimates a fundamental matrix $F'$ very well, when using B as input. We have then that for every pair of homologous points $(x, \hat{x}) \in A$, this expression is valid

$$\hat{x}^T (H_2^T F H_1) x = (H_2 \hat{x})^T F (H_1 x) = 0.$$

This shows that the fundamental matrix that establishes the epipolarity of the points of A is defined by $F = H_2^T F' H_1$.

## 5.8 FINDING THE EXTRINSIC PARAMETERS

If the intrinsic parameters of a camera are known, it is possible, from of a fundamental matrix $F$, to determine the possible relative positions between two cameras that explain this fundamental matrix.

Given the fundamental matrix $F = K_2^{-T} [t]_\times R K^{-1}$ establishing the relationship of epipolarity of projections obtained by the cameras $K_1[I|0]$ and $K_2[R|t]$, we can define an essential matrix

$$E = K_2^T F K_1, \tag{5.4}$$

which relates the projections captured by the cameras $[I|0]$ and $[R|t]$

The matrix $E = [t]_\times R$ is the product of the anti-symmetric matrix $[t]_\times$ by rotation matrix R. The determination of the possible values of t and R is solved by the theorem below, whose demonstration can be found in [13].

**Theorem 5.3.** *Assuming that the SVD decomposition of an essential matrix $E$ is equal to $Udiag(1, 1.0)V^T$, there are two ways to factor $E$, so that $E = SR$, such that $S$ is an anti-symmetric matrix and $R$ is a rotation matrix. We have that $S = UZU^T$, and $R = UWV^T$ or $R = UW^T V^T$, in which*

$$W = \begin{pmatrix} 0 & -1 & 0 \\ 1 & 0 & 0 \\ 0 & 0 & 1 \end{pmatrix}$$

*and*

$$Z = \begin{pmatrix} 0 & 1 & 0 \\ -1 & 0 & 0 \\ 0 & 0 & 0 \end{pmatrix}.$$

The Theorem 5.3 shows two possible choices for the rotation matrix $R$. To determine which are the possible vectors t, just take into account the following facts:

1. $[t]_\times t = t \times t = 0$.

2. The vector is defined with a scale ambiguity.

Using the notation of the theorem, we have for the first fact that, every vector $t$ must belong to the nucleus of $[t]_\times$. Since $[t]_\times = S = UZU^T$, we concluded that all vector t must be of the form

$$t = \lambda U(0,0,1)^T,$$

such that $\lambda \in \mathbb{R}$.

The second fact demonstrated in Section 5.1 implies that $t$ can be any element of the form $\lambda U(0,0,1)^T$ , with $\lambda \in \mathbb{R}$. If we restrict ourselves to cases where $\|t\| = 1$, we have that $t$ can be $U(0,0,1)^T$ or $-U(0,0,1)^T$.

### 5.8.1 Adding Clipping to the Model

We can conclude from the previous section that, being known a fundamental matrix $F$, which relates projections obtained by a pair of cameras $P_1$ and $P_2$, whose intrinsic parameters are defined by $K_1$ and $K_2$ matrices . If $P_1 = K_1[I|0]$ then $P_2$ can be defined from four ways:

$$K_2[UWV^T]U(0,0,1)^T],$$
$$K_2[UW^TV^T|U(0,0,1)^T],$$
$$K_2[UWV^T|-U(0,0,1)^T],$$
$$K_2[UW^TV^T|-U(0,0,1)^T],$$

in which $U$ and $W$ can be calculated from $F$ using Equation 5.4 and the Theorem 5.3.

By stating that there are only these four solutions, we are considering that the indetermination of the distance between the projection centers of $P_1$ and $P_2$ is implicit, such as explained in Section 5.1.

The camera model we are using does not describe the clipping operation in relation to the vision pyramid. The result of this is that, only one of these four camera configurations are physically achievable, as exemplified by Figure 5.1.

The solution to this problem is to discard the solutions that make that the three-dimensional reconstruction of homologous points has a negative z coordinate for any of the references defined by the cameras [13]. We will show how to get three-dimensional reconstruction of a point from its projections in the next section.

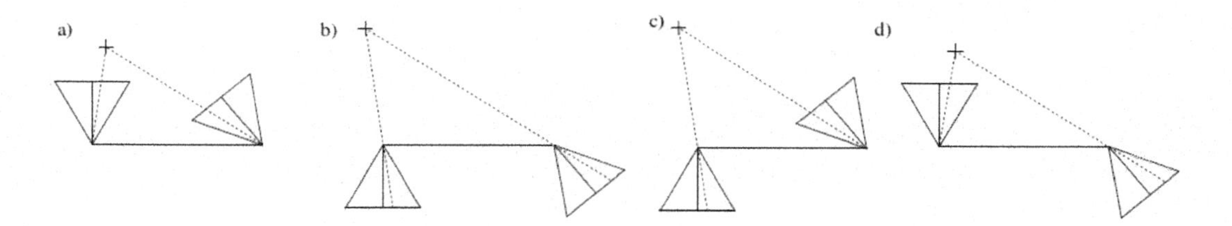


Figure 5.1 Although there are four configurations that projectively explain the pair of homologous points, only in (a) the projected point is positioned in front of both the cameras.

### 5.8.2 Three-Dimensional Reconstruction

Let $x_1 \in \mathbb{R}P^2$ and $x_2 \in \mathbb{R}P^2$ the projections of a point $X \in \mathbb{R}P^3$ on the cameras $P_1$ and $P_2$ , that is, $x_1 = P_1X$ and $x_2 = P_2X$. We will now present how to determine $X$ when $x_1$, $x_2$, $P_1$ and $P_2$ are known.

Interpreting $x_1 = (u, v, 1)^T$ and $P_1X$ as vectors of $\mathbb{R}^3$ , we have $x_1 \times (P_1X) = 0$. Calling $P_i^n$ the nth line of $P_i$, this expression can be rewritten as the following set of three linear equations in X, where two are linearly independent:

$$u(P_1^3X) - (P_1^1X) = 0,$$
$$v(P_1^3X) - (P_1^2X) = 0,$$
$$u(P_1^2X) - v(P_1^1X) = 0,$$

Similarly, $x_2 = (u, v, 1)^T$ can be used to get more two other linear equations in $X$, and linearly independent, that $x_2 \times (P_2X) = 0$. Joing four of these equations we get a linear system homogeneous form $AX = 0$, such that

$$A = \begin{pmatrix} uP_1^3 - P_1^1 \\ vP_1^3 - P_1^2 \\ u'P_2^3 - P_2^1 \\ v'P_2^3 - P_2^2 \end{pmatrix}.$$

This is a linear system of four equations over the four homogeneous coordinates of $X$, therefore it is a overconstrained linear system, which can be converted as the optimization problem $\min_{\|X\|=1} \|AX\|$, whose solution is given by Theoremn 3.2.

## 5.9 API

```
void calib_fmatrix_make( gsl_matrix *f, gsl_matrix *b,
                         gsl_matrix *a );
```

This function implements the algorithm of Section 5.6. $b$ and $a$ encodes in each line a pair of homologous 2D points, and $f$ returns the related Fundamental Matrix.

```
void calib_fmatrix_singular_enforce( gsl_matrix *r,
                                     gsl_matrix *f );
```

This function implements the singular enforcement of the Fundamental Matrix described in the Theorem 5.2. $r$ is the closest singular fundamental matrix to the fundamental matrix $f$.

This is used by the function *calib_fmatrix_dlt* since the function *calib_fmatrix_make* can returns in general a non-singular matrix.

```
void calib_fmatrix_dlt( gsl_matrix *f, gsl_matrix *b,
                        gsl_matrix *a );
```

This function implements the algorithm of Section 5.7. $b$ and $a$ encodes in each line a pair of homologous 2D points, and $f$ returns the related Fundamental Matrix.

```
void calib_ematrix_make( gsl_matrix *e, gsl_matrix *f,
                         gsl_matrix *k );
```

This function extracts an Essential Matrix $e$ from the Fundamental Matrix $f$ and intrinsic parameters $k$ for both cameras, applying the Equation 5.4.

```
void calib_ematrix_get_all_rt( gsl_matrix *q,
     gsl_matrix *r1, gsl_matrix *r2, gsl_vector *t );
```

This function is used to find all the possible pair of extrinsic parameters: $[r1|t]$, $[r1|-t]$, $[r2|t]$, $[r2|-t]$ related to the Essential Matrix received by the vector $q$. It implements the theory explained by Section 5.8.1.

```
int is_rotation( gsl_matrix *r );
```

It is an auxiliary function used by the function *calib_ematrix_get_all_rt* which returns 1 if the elements of the orthogonal matrix $r$ is a rotation matrix, and it returns 0 otherwise.

```
void calib_get_3dpoint( gsl_matrix *p1, gsl_matrix *p2,
         gsl_vector *x1, gsl_vector *x2, gsl_vector *x )
```

This function implements the algorithm described in the Section 5.8.2. It receives two $3 \times 4$ matrices $p1$ and $p2$, related to two cameras, and two homologous points $x1$ and $x2$. Then it returns on $x$ the 3D reconstruction of them.

```
void calib_ematrix_get_RT( gsl_matrix *k, gsl_matrix *e,
gsl_vector *x1, gsl_vector *x2, gsl_matrix *r, gsl_vector
*tr )
```

This function fully implements the algorithm of Section 5.8. $k$ receives the correspondent intrinsic parameter of the cameras, $e$ receives the Essential Matrix. $x1$ and $x2$ receive a pair of homologous points. $r$ returns the rotation of the camera, and $tr$ returns the correspondent translation.

```
char infront( gsl_matrix *pid, gsl_matrix *k, gsl_matrix *r,
              gsl_vector *t, gsl_vector *x1, gsl_vector *x2 )
```

This is an auxiliary function used by the function *calib_ematrix_get_RT*. *pid* must receives the camera $k[I|0]$, the second camera is $k[r|t]$ encoded in the respective parameters, and $x1$ and $x2$ are two homologous points. If the 3D reconstruction of $x1$ and $x2$ is in front of both cameras the function returns 1, otherwise it returns 0.

```
void calib_rt_transform( gsl_vector *xcam, gsl_matrix *r,
                         gsl_vector *t, gsl_vector *x )
```

This function changes the vector $x$ assuming a new referential system specified by the rotation matrix $r$ and by the translation vector $t$. The result is returned in $xcam$.

```
void calib_ematrix_get_P( gsl_matrix *k, gsl_matrix *e,
          gsl_vector *x1, gsl_vector *x2, gsl_matrix *p )
```

This function receives the intrinsic parameter of a camera in the parameter $k$, an Essential Matrix $e$, and receives a pair of homologous points $x1$ and $x2$ and returns one correspondent camera in the matrix $p$ satisfying the fact that the reconstruction of $x1$ and $x2$ is in front the pair of cameras. The other matrix related to the other camera is $k[I|0]$ such as explained in Section 5.8.

```
void calib_dlt_reconstruct( gsl_matrix *points,
    gsl_matrix *projs1, gsl_matrix *projs2,  gsl_matrix *k  )
```

This function combines many function described previously in order to find a reconstruction of a set of homologous points, assuming that the intrinsic parameter of the camera used to capture them are known. $k$ receives the intrinsic parameters, $projs1$ and $projs2$ receive both a set of homologous pairs encoded in the lines of the matrices. The result is a $n \times 3$ matrix *point* whose lines encodes the 3D vector related to each pair of homologous points.

## 5.10 CODE

```
// calib/fmatrix.c

/* 8-points algorithm */
void calib_fmatrix_make( gsl_matrix *f ,gsl_matrix *b, gsl_matrix *a )
{
 int i, n;
 gsl_matrix *u, *v;
 gsl_vector *s, *x, *w;

 n = b->size1;
 u = gsl_matrix_alloc( n, 9 );
 v = gsl_matrix_alloc( 9, 9 );
 s = gsl_vector_alloc( 9 );
 x = gsl_vector_alloc( 9 );
 w = gsl_vector_alloc( 9 );

 for( i=0; i<n; i++ ){
   gsl_matrix_set( u, i, 0, gsl_matrix_get(a,i,0)*gsl_matrix_get(b,i,0) );
   gsl_matrix_set( u, i, 1, gsl_matrix_get(a,i,0)*gsl_matrix_get(b,i,1) );
   gsl_matrix_set( u, i, 2, gsl_matrix_get(a,i,0) );
   gsl_matrix_set( u, i, 3, gsl_matrix_get(a,i,1)*gsl_matrix_get(b,i,0) );
   gsl_matrix_set( u, i, 4, gsl_matrix_get(a,i,1)*gsl_matrix_get(b,i,1) );
   gsl_matrix_set( u, i, 5, gsl_matrix_get(a,i,1) );
   gsl_matrix_set( u, i, 6, gsl_matrix_get(b,i,0) );
   gsl_matrix_set( u, i, 7, gsl_matrix_get(b,i,1) );
   gsl_matrix_set( u, i, 8, 1. );
 }

 gsl_linalg_SV_decomp( u, v, s, w );
 gsl_matrix_get_col( x, v, 8);

 for( i=0; i<9; i++ )
   gsl_matrix_set( f, i/3 , i%3 , gsl_vector_get( x,i ));

 gsl_matrix_free( u );
 gsl_matrix_free( v );
 gsl_vector_free( s );
 gsl_vector_free( x );
 gsl_vector_free( w );
}


/* singular enforcement */
void calib_fmatrix_singular_enforce( gsl_matrix *r, gsl_matrix *f )
{
 gsl_vector *s, *w;
 gsl_matrix *u, *v, *sm, *usm;

 u = gsl_matrix_alloc(3,3);
 v = gsl_matrix_alloc(3,3);
 s = gsl_vector_alloc(3);
 sm = gsl_matrix_alloc(3,3);
 usm = gsl_matrix_alloc(3,3);
 w = gsl_vector_alloc(3);

 gsl_matrix_memcpy( u, f );
 gsl_linalg_SV_decomp( u, v, s, w );
 gsl_matrix_set_zero( sm );
 gsl_matrix_set( sm, 0, 0, gsl_vector_get( s, 0 ) );
 gsl_matrix_set( sm, 1, 1, gsl_vector_get( s, 1 ) );
 gsl_matrix_set( sm, 2, 2, 0. );
```

```
 gsl_linalg_matmult( u , sm , usm );
 gsl_matrix_transpose( v );
 gsl_linalg_matmult( usm, v, r );

 gsl_matrix_free(u);
 gsl_matrix_free(v);
 gsl_vector_free(s);
 gsl_matrix_free(sm);
 gsl_matrix_free(usm);
 gsl_vector_free(w);
}


void calib_fmatrix_dlt( gsl_matrix *f, gsl_matrix *b, gsl_matrix *a )
{
 gsl_matrix *h1, *h2, *b2, *a2, *f1, *f1_h1, *f1_sing, *f1_sing_h1;

 h1 = gsl_matrix_alloc( 3, 3 );
 h2 = gsl_matrix_alloc( 3, 3 );
 b2 = gsl_matrix_alloc( b->size1, 2 );
 a2 = gsl_matrix_alloc( a->size1, 2 );
 f1 = gsl_matrix_alloc( 3, 3 );
 f1_sing = gsl_matrix_alloc( 3, 3 );
 f1_h1 = gsl_matrix_alloc( 3, 3 );
 f1_sing_h1 = gsl_matrix_alloc( 3, 3 );

 calib_get_normalize_transform( h1, NULL, b );
 calib_get_normalize_transform( h2, NULL, a );
 calib_apply_homog( b2, h1, b );
 calib_apply_homog( a2, h2, a );

 calib_fmatrix_make( f1, b2, a2 );
 calib_fmatrix_singular_enforce( f1_sing, f1 );
 gsl_linalg_matmult( f1_sing, h1, f1_sing_h1 );
 gsl_matrix_transpose( h2 );
 gsl_linalg_matmult( h2, f1_sing_h1, f );

 gsl_matrix_free( h1 );
 gsl_matrix_free( h2 );
 gsl_matrix_free( b2 );
 gsl_matrix_free( a2 );
 gsl_matrix_free( f1 );
 gsl_matrix_free( f1_h1 );
 gsl_matrix_free( f1_sing );
 gsl_matrix_free( f1_sing_h1 );
}

// calib/ematrix.c

void calib_ematrix_make( gsl_matrix *e, gsl_matrix *f, gsl_matrix *k )
{
 gsl_matrix *aux, *kaux;

 aux = gsl_matrix_alloc( 3, 3 );
 kaux = gsl_matrix_alloc( 3, 3 );

 gsl_linalg_matmult( f, k, aux );
 gsl_matrix_memcpy( kaux, k );
 gsl_matrix_transpose( kaux );
 gsl_linalg_matmult( kaux, aux, e );

 gsl_matrix_free( aux );
 gsl_matrix_free( kaux );
```

```
}


void calib_ematrix_get_all_rt( gsl_matrix *q, gsl_matrix *r1,
                               gsl_matrix *r2, gsl_vector *t )
{
 gsl_matrix *e, *v, *aux, *qaux;
 gsl_vector *s, *w;

 s = gsl_vector_alloc( 3 );
 w = gsl_vector_alloc( 3 );
 v = gsl_matrix_alloc( 3, 3 );
 e = gsl_matrix_alloc( 3, 3 );
 aux = gsl_matrix_alloc( 3, 3 );
 qaux = gsl_matrix_alloc( 3, 3 );
 gsl_matrix_memcpy( qaux, q );

 /* creating matrix W */
 gsl_matrix_set_zero(e);
 gsl_matrix_set( e, 0, 1,  1. );
 gsl_matrix_set( e, 1, 0, -1. );
 gsl_matrix_set( e, 2, 2,  1. );

 gsl_linalg_SV_decomp( qaux, v, s, w );
 gsl_linalg_matmult( qaux, e, aux );
 gsl_matrix_transpose( v );
 gsl_linalg_matmult( aux, v, r1 );
 if( !is_rotation(r1) )
   gsl_matrix_scale( r1, -1. );
 gsl_matrix_get_col( t, qaux, 2 );

 gsl_matrix_transpose(e);
 gsl_linalg_matmult( qaux, e, aux );
 gsl_linalg_matmult( aux, v, r2 );
 if( !is_rotation(r2) )
   gsl_matrix_scale( r2, -1. );

 gsl_vector_free(s);
 gsl_vector_free(w);
 gsl_matrix_free(v);
 gsl_matrix_free(e);
 gsl_matrix_free(aux);
 gsl_matrix_free(qaux);
}


/* Return 1 if <T(i)XT(j), T(K)> > 0 */
int is_rotation( gsl_matrix *r )
{
 double r11, r12, r13,
        r21, r22, r23,
        r31, r32, r33;

 r11 = gsl_matrix_get(r, 0, 0);
 r12 = gsl_matrix_get(r, 0, 1);
 r13 = gsl_matrix_get(r, 0, 2);
 r21 = gsl_matrix_get(r, 1, 0);
 r22 = gsl_matrix_get(r, 1, 1);
 r23 = gsl_matrix_get(r, 1, 2);
 r31 = gsl_matrix_get(r, 2, 0);
 r32 = gsl_matrix_get(r, 2, 1);
 r33 = gsl_matrix_get(r, 2, 2);

 if( r13*( r21*r32 - r22*r31 ) - r23*(r11*r32 - r31*r12)
```

```
        + r33*( r11*r22 - r12*r21 ) > 0 )
      return 1;
   else
        return 0;
}

// calib/get3d.c

#define CREATE_ROW( R, P, X, J )  \
 gsl_matrix_get_row( aux1, P, 2 ); \
 gsl_vector_scale( aux1, gsl_vector_get( X, J ) ); \
 gsl_matrix_get_row( aux2, P, J ); \
 gsl_vector_sub( aux1, aux2 ); \
 gsl_matrix_set_row( a, R, aux1 );

void calib_get_3dpoint( gsl_matrix *p1, gsl_matrix *p2,
                        gsl_vector *x1, gsl_vector *x2, gsl_vector *x )
{
 gsl_vector *aux1, *aux2, *s, *w, *r;
 gsl_matrix *a, *v;

 a = gsl_matrix_alloc( 4, 4 );
 v = gsl_matrix_alloc( 4, 4 );
 aux1 = gsl_vector_alloc( 4 );
 aux2 = gsl_vector_alloc( 4 );
 s = gsl_vector_alloc( 4 );
 w = gsl_vector_alloc( 4 );
 r = gsl_vector_alloc( 4 );

 CREATE_ROW( 0, p1, x1, 0 );
 CREATE_ROW( 1, p1, x1, 1 );
 CREATE_ROW( 2, p2, x2, 0 );
 CREATE_ROW( 3, p2, x2, 1 );

 gsl_linalg_SV_decomp( a, v, s, w );
 gsl_matrix_get_col( r, v, 3 );

 gsl_vector_set( x, 0, gsl_vector_get(r,0)/gsl_vector_get(r,3));
 gsl_vector_set( x, 1, gsl_vector_get(r,1)/gsl_vector_get(r,3));
 gsl_vector_set( x, 2, gsl_vector_get(r,2)/gsl_vector_get(r,3));

 gsl_matrix_free( a );
 gsl_matrix_free( v );
 gsl_vector_free( aux1 );
 gsl_vector_free( aux2 );
 gsl_vector_free( s );
 gsl_vector_free( w );
 gsl_vector_free( r );
}


// calib/getRT.c

char infront( gsl_matrix *pid, gsl_matrix *k, gsl_matrix *r,
              gsl_vector *t, gsl_vector *x1, gsl_vector *x2 );

#define COPY_RT( R, RSRC, T, TSRC )  { gsl_matrix_memcpy( R, RSRC ); \
                                       gsl_vector_memcpy( T, TSRC ); }

void calib_ematrix_get_RT( gsl_matrix *k, gsl_matrix *e, gsl_vector *x1,
                           gsl_vector *x2, gsl_matrix *r, gsl_vector *tr )
{
 gsl_vector *t;
 gsl_matrix *r1, *r2, *id, *pid;
```

```
 r1 = gsl_matrix_alloc(3,3);
 r2 = gsl_matrix_alloc(3,3);
 id = gsl_matrix_alloc(3,3);
 pid = gsl_matrix_alloc(3,4);
 t = gsl_vector_alloc(3);

 gsl_vector_set_zero(t);
 gsl_matrix_set_zero(id);
 gsl_matrix_set( id, 0,0,1.);
 gsl_matrix_set( id, 1,1,1.);
 gsl_matrix_set( id, 2,2,1.);

 calib_pmatrix_make( pid, k, id, t );
 calib_ematrix_get_all_rt( e, r1, r2, t );

 if( infront( pid, k, r1, t, x1, x2 ) )
   COPY_RT( r, r1, tr, t )
 if( infront( pid, k, r2, t, x1, x2 ) )
   COPY_RT( r, r2, tr, t )
 gsl_vector_scale( t, -1. );
 if( infront( pid, k, r1, t, x1, x2 ) )
   COPY_RT( r, r1, tr, t )
 if( infront( pid, k, r2, t, x1, x2 ) )
   COPY_RT( r, r2, tr, t )

 gsl_matrix_free(r1);
 gsl_matrix_free(r2);
 gsl_matrix_free(id);
 gsl_matrix_free(pid);
 gsl_vector_free(t);
}


char infront( gsl_matrix *pid, gsl_matrix *k, gsl_matrix *r,
              gsl_vector *t, gsl_vector *x1, gsl_vector *x2 )
{
 gsl_vector *x,*xcam;
 gsl_matrix *p;

 x = gsl_vector_alloc(3);
 p = gsl_matrix_alloc(3, 4);
 calib_pmatrix_make( p, k, r, t );
 calib_get_3dpoint( pid, p, x1, x2, x );
 gsl_matrix_free(p);

 if( gsl_vector_get(x, 2) < 0 ){
    gsl_vector_free(x);
    return 0;
 }
 else{
    gsl_vector *xcam;
    xcam = gsl_vector_alloc(3);
    calib_rt_transform( xcam, r, t, x );
    if( gsl_vector_get(xcam, 2) < 0 ){
       gsl_vector_free(x); gsl_vector_free(xcam);
       return 0;
    }
 }
 return 1;
}


// calib/rttransform.c
```

```
#include "calib.h"

void calib_rt_transform( gsl_vector *xcam, gsl_matrix *r, gsl_vector *t,
                         gsl_vector *x )
{
 gsl_matrix *aux1, *aux2;

 aux1 = gsl_matrix_alloc(3, 1);
 aux2 = gsl_matrix_alloc(3, 1);
 gsl_matrix_set_col(aux1, 0, x);
 gsl_linalg_matmult(r, aux1, aux2);
 gsl_matrix_set_col(aux1, 0, t);
 gsl_matrix_add(aux1, aux2);
 gsl_matrix_get_col(xcam, aux1, 0);

 gsl_matrix_free(aux1);
 gsl_matrix_free(aux2);
}


// calib/getP.c

void calib_ematrix_get_P( gsl_matrix *k, gsl_matrix *e, gsl_vector *x1,
                          gsl_vector *x2, gsl_matrix *p )
{
 gsl_vector *t;
 gsl_matrix *r;

 t = gsl_vector_alloc(3);
 r = gsl_matrix_alloc(3,3);

 calib_ematrix_get_RT( k, e, x1, x2, r, t );
 calib_pmatrix_make( p, k, r, t );

 gsl_matrix_free(r);
 gsl_vector_free(t);
}


// calib/reconstruct.c

void calib_dlt_reconstruct(  gsl_matrix *points, gsl_matrix *projs1,
                             gsl_matrix *projs2,  gsl_matrix *k  )
{
 int i;
 gsl_matrix *f, *e, *p, *pid, *id;
 gsl_vector *zero, *x1, *x2, *x;

 f = gsl_matrix_alloc( 3, 3 );
 e = gsl_matrix_alloc( 3, 3 );
 p = gsl_matrix_alloc( 3, 4 );
 pid = gsl_matrix_alloc( 3, 4 );
 id = gsl_matrix_alloc( 3, 3 );
 zero = gsl_vector_alloc( 3 );
 x1 = gsl_vector_alloc( 2 );
 x2 = gsl_vector_alloc( 2 );
 x = gsl_vector_alloc( 3 );

 gsl_matrix_set_identity( id );
 gsl_vector_set_zero( zero );

 calib_fmatrix_make(  f, projs1 , projs2 );
 calib_ematrix_make( e, f, k );
```

```
 gsl_matrix_get_row( x1, projs1, 0 );
 gsl_matrix_get_row( x2, projs2, 0 );
 calib_ematrix_get_P( k, e, x1, x2, p );
 calib_pmatrix_make( pid, k, id, zero );

 for( i = 0; i < projs1->size1; i++ ) {
    gsl_matrix_get_row( x1, projs1, i );
    gsl_matrix_get_row( x2, projs2, i );
    calib_get_3dpoint( pid, p, x1, x2, x  );
    gsl_matrix_set( points, i, 0, gsl_vector_get( x, 0 ) );
    gsl_matrix_set( points, i, 1, gsl_vector_get( x, 1 ) );
    gsl_matrix_set( points, i, 2, gsl_vector_get( x, 2 ) );
 }

 gsl_matrix_free( f );
 gsl_matrix_free( e );
 gsl_matrix_free( p );
 gsl_matrix_free( pid );
 gsl_matrix_free( id );
 gsl_vector_free( zero );
 gsl_vector_free( x1 );
 gsl_vector_free( x2 );
 gsl_vector_free( x );
}
```

CHAPTER 6

# Feature Tracking

THE NEXT CHAPTER will present a calibration process for family of cameras. Such a process needs the correspondence between projections of different points of the scene over several frames of a video. Keeping in mind that even short-duration videos are made up of hundreds of frames, it is necessary that this correspondence is made automatically. We will describe in this chapter the algorithm Kanade-Lucas-Tomasi, which will be used to solve this problem. The description more detailed information can be found in [19] and [28]

## 6.1 DEFINITIONS

We will adopt the following definitions:

1. **Image**

   An image is a function $I : [a,b] \times [c,d] \to \mathbb{R}$. In this case, we are considering a model for grayscale images, where for each support point $[a,b] \times [c,d]$ a brightness value is associated.

2. **Video**

   A video is a finite family of images $(I)_n = (I_1, ..., I_n)$, where each image $I_k$ corresponds to a frame captured by a camera. Also, the order defined by the indexing of the tables corresponds to the order in which the frames were captured by the camera.

3. **Window**

   A window of an image: $I : [a,b] \times [c,d] \to \mathbb{R}$ is an $I|_w$ obtained by restriction of the domain of $I$ to a small rectangle $w = [a',b'] \times [c',d'] \subset [a,b] \times [c,d]$.

## 6.2 KANADE-LUCAS-TOMASI ALGORITHM

Kanade-Lucas-Tomasi (KLT) is an algorithm capable of tracking windows in the video. Given a video $(I)_n$ , it seeks to locate windows in an $I_{j+1}$ frame that correlated by a translation with windows of $I_j$ . More precisely, the algorithm is able to determine

DOI: 10.1201/9781003206026-6

a vector $h \in \mathbb{R}^2$ , called disparity, such that

$$\forall x \in w, I_{j+1}|_{w'}(x+h) = I_j|_w(x) + \eta(x) \tag{6.1}$$

in which $w'$ is the rectangle obtained by adding $h$ to the vertices of $w$, and $\eta : w \to \mathbb{R}^+$ is a function that defines the point correlation error between the windows. The algorithm then searches determine the disparity $h$ that minimizes this error over the entire window.

The usefulness of correlating windows for our purposes is the fact that windows that are similar, and are close together in consecutive frames, have a great chance of corresponding to the projection of the same set of points in the three-dimensional scene. This means that, since $x_0 \in w$, it is reasonable to use $x_0 + h$ as an estimate for its homologous point in $I_{j+1}$. In the calibration process presented in the next chapter, the window centers correlated by KLT are used as homologous points. In this way, the point tracking problem is converted in a window tracking problem.

## 6.3 FOLLOWING WINDOWS

Using the notation established in equation 6.1, and having fixed a vector disparity $h$, we can define a measure for the correlation error as

$$E = \int_w \eta(x)^2 \, dx \tag{6.2}$$

In this way, the problem of determining the disparity can be formalized through the following optimization problem:

**Problem 6.3.1.** *Find a vector $h \in \mathbb{R}^2$ that minimizes $\int_w [I_{j+1}(x+h) - I_j(x)]^2 \, dx$, in which $w$ is the rectangle that defines the window in $I_j$.*

Performing the change of variables $\nu = x + h$, we have that this problem is equivalent to finding the vector $h \in \mathbb{R}^2$ that minimizes

$$\int_w [I_{j+1}(\nu) - I_j(\nu - h)]^2 \, dx,$$

assuming that $I_{j+1}$ is differentiable, and that the disparity between consecutive frames is small, we can make the following approximation

$$I_j(\nu - h) \approx I_j(\nu) - I_j'(\nu) \cdot h.$$

By doing that, we have that the objective function can be rewritten as

$$h \mapsto \int_w [\Phi(\nu) - I_j'(\nu) \cdot h]^2 \, d\nu, \text{ in which } \Phi(\nu) = I_{j+1}(\nu) - I_j(\nu).$$

This function has a minimum at a critical point $h = (h_1, h_2)^T$ satisfying

$$\forall u \in \mathbb{R}^2, \int_w [\Phi(\nu) - I_j'(\nu) \cdot h](I_j' \cdot u) d\nu = 0.$$

In particular, this property is valid when we assume that $u$ is the canonical base vectors $(1,0)^T$ or $(0,1)^T$, allowing us to rewrite the above expression in terms of derivatives partial, obtaining the following linear system, which allows us to determine $h$ :

$$\left[\int_w \left(\frac{\partial I_j}{\partial x_1}(\nu)\right)^2 d\nu\right] h_1 + \left[\int_w \frac{\partial I_j}{\partial x_1}(\nu)\frac{\partial I_j}{\partial x_2}(\nu)d\nu\right] h_2 = \int_w \Phi(\nu)\frac{\partial I_j}{\partial x_1}(\nu)d\nu$$

$$\left[\int_w \left(\frac{\partial I_j}{\partial x_2}(\nu)\right)^2 d\nu\right] h_2 + \left[\int_w \frac{\partial I_j}{\partial x_1}(\nu)\frac{\partial I_j}{\partial x_2}(\nu)d\nu\right] h_1 = \int_w \Phi(\nu)\frac{\partial I_j}{\partial x_2}(\nu)d\nu$$

## 6.4 CHOOSING THE WINDOWS

In addition to defining a window tracking algorithm, the algorithm KLT defines an automatic process for selecting windows to be monitored. This selection process is based on a criterion defined so that the solution of the linear system defined in the previous section can be obtained with high precision.

Consider the system written in the form $Ah = b$. In order to achieve an accurately solution for this system, it is necessary that it is well conditioned, and that the coefficients of matrix $A$ is defined above the noise level of the image.

The system is well conditioned if the two eigenvalues of $A$, $\lambda_1$ and $\lambda_2$, are of the same order of magnitude. In practice, this always happens, since the brightness value at each point of the image is limited.

In order to get coefficients of $A$ defined above the noise level of the image it is necessary that $\lambda_1$ and $\lambda_2$ are not small. Therefore, the KLT algorithm uses $min\{\lambda_1, \lambda_2\}$ as a quality measure for tracking a window chosen.

The choice of the $m$ windows in frame $I_k : U \to R$ that are best tracked is done by comparing the values of quality of all possible choices of windows $w \subset U$. The chosen windows are the ones with the better quality, and that are delimited by rectangles $w_1, \ldots, w_m$ that do not overlap, that is, $w_i \bigcap w_j = \emptyset$, for $i, j \in \{1, \ldots, n\}$.

Another option, used for example in the OpenCV library, consists on establishing a minimum distance between the chosen windows.

## 6.5 DISPOSAL OF WINDOWS

After determining the disparity vector, the algorithm evaluates the correlation error. defined in Equation 6.2. If that value exceeds a certain threshold, it stops accompany the window from that frame, because the disparity obtained against windows that are very different. This window can be replaced by a new one, which must be chosen as the best trackable in the frame, and that does not overlap other windows that are still being tracked.

## 6.6 PROBLEMS USING KLT

Our interest in the KLT algorithm is to use it to determine the projections of a set of points of a three-dimensional scene in a video. Unfortunately there are no guarantees that the projections found by it satisfy this property.

One of the problems is that the disposal strategy of the KLT algorithm avoids only large errors made in consecutive frames. It does not prevent the accumulation of small errors during an accompaniment on a sequence of frames. This means that the results may be inaccurate, especially in the case of tracking points over long sequences of frames.

Another problem can be understood by analyzing Figure 6.1, which shows three frames of a video in which the KLT algorithm was applied to follow the projection of 20 points of the scene.

Figure 6.1 Example of points that are not fixed points in the 3D scene. In the case of point 1 KLT is tracking a region of brightness on a surface. The problem is that this region moves with the movement of the camera. In the case of the point 2, the KLT is tracking the superposition of the projection of the edges of two surfaces distinct from the scene.

It is clearly seen that the two points indicated in the images are problematic, since they do not correspond to the projections of fixed points in the scene. A modification of the KLT algorithm that seeks to solve the first problem can be found in [25]. In this modification, in addition to tracking points in consecutive frames, the comparison of the neighborhood of each point is made with the neighborhood of your correspondent in the frame in which he was selected to be tracked. If the neighborhoods become very different the point is no longer tracked. This version of KLT was not used in this book, because of the hypothesis rigidity of the scene allows us to solve both the first and the second problem simultaneously. One way to do this will be presented in the next chapter.

## 6.7 CODE

We did not implement the KLT library. Instead, we decided to use the KLT algorithm available in the OpenCV library.

This following code implements a program that write, in *stdout*, the coordinates list of tracked points in two columns. After each pair of coordinates, there are a number "0" if the tracking was a successful and "−1" if it fails to track.

The user must pass two parameters to the program: $argv[1]$ is the name of the video file, $argv[2]$ is the number of features to be tracked, and $argv[3]$ is the minimum distance between tracked features.

```
// track/main.cxx

#include <iostream>
#include <opencv2/core.hpp>
#include <opencv2/highgui.hpp>
#include <opencv2/imgproc.hpp>
#include <opencv2/videoio.hpp>
#include <opencv2/video.hpp>

using namespace cv;
using namespace std;

#define SQR(X) ((X)*(X))


static double min_distancy( Point2f p2, vector<Point2f> p1 );

int main(int argc, char **argv)
{
 VideoCapture capture(argv[1]);

  Mat old_frame, old_gray;
  vector<Point2f> p0, p1, p2;
  int npoints = atoi(argv[2]);
  double features_dist = atof(argv[3]);

  capture >> old_frame;
  cvtColor(old_frame, old_gray, COLOR_BGR2GRAY);
  goodFeaturesToTrack(old_gray, p0, npoints, 0.00001, features_dist );
  std::cout << "Features per frame = " << argv[2] << std::endl;
  std::cout << "Frame 0"  << std::endl;
  for(uint i = 0; i < p0.size(); i++)
      std::cout << p0[i].x << " " << p0[i].y << " 0"  << std::endl;
  Mat mask = Mat::zeros(old_frame.size(), old_frame.type());
  int k = 1;
  while(true){
      Mat frame, frame_gray;
      capture >> frame;
      if (frame.empty())
            break;
      std::cout << "Frame " << k++ << std::endl;
      cvtColor(frame, frame_gray, COLOR_BGR2GRAY);
        vector<uchar> status;
        vector<float> err;
        TermCriteria criteria = TermCriteria((TermCriteria::COUNT)
                                + (TermCriteria::EPS), 30, 0.01);
        calcOpticalFlowPyrLK(old_gray, frame_gray, p0, p1, status, err );
        vector<Point2f> good_new;
        for(uint i = 0; i < p0.size(); i++)
        {
            if(status[i] == 1) {
                good_new.push_back(p1[i]);
                circle(frame, p1[i], 2, Scalar(0,0,255), -1);
                std::cout << p1[i].x << " " << p1[i].y << " 0"  << std::endl;
            }
            else {
                std::cout << p1[i].x << " " << p1[i].y << " -1"  << std::endl;
                int n = 0;
```

```
                do{
                  goodFeaturesToTrack(old_gray, p2, n+1, 0.00001, 4 );
                  if( (min_distancy( p2[n], good_new ) > features_dist) &&
                      (min_distancy( p2[n], p1 ) > features_dist) ){
                         good_new.push_back( p2[n] );
                         n = -1;
                  }
                  n++;
                }
                while( n != 0 );
            }
        }
        Mat img;
        add(frame, mask, img);
        imshow("Frame", img);
        int keyboard = waitKey(30);
        if (keyboard == 'q' || keyboard == 27)
            break;
        old_gray = frame_gray.clone();
        p0 = good_new;
    }
}


static double min_distancy( Point2f p2, vector<Point2f> p1 )
{
 Point2f p;
 double dist, min_dist = 1000000;

 for( int i = 0; i < p1.size(); i++ ){
   dist = SQR(p2.x - p1[i].x) + SQR(p2.y - p1[i].y);
   if( dist < min_dist )
      min_dist = dist;
 }
 return min_dist;
}
```

CHAPTER 7

# Estimating Many Cameras

IN THIS CHAPTER we describe a robust algorithm, capable of determining the extrinsic parameters assumed by a camera to capture the frames of a video, given that the intrinsic parameters were previously estimated, using what was seen in Chapter 4. The scene that is presented in the video needs to be predominantly rigid, that is, most of the points in the scene cannot have their position changed, because the constraints imposed by the rigidity on its projections make it possible to determine camera parameters.

## 7.1 DEFINITIONS

We will adopt the following definitions:

1. **Family of homologous points**

   Given a video $(I)_n = (I_1, \ldots, I_n)$, we say that the family $(x)_n = (x_1, \ldots, x_n)$, in which $x_i \in \mathbb{R}P^2$ , is a family of homologous points associated with video $(I)_n$ if there is a point $X \in \mathbb{R}P^3$ , of the scene, such that the projection of $X$ in $I_j$ is $x_j$ , for all $j \in \{1, \ldots, n\}$.

2. **Matrix of homologous points**

   The matrix $M$, $m \times n$, formed by elements of $\mathbb{R}P^2$ , is a matrix of points homologous associated with a video $(I)_n$ if each of its lines defines a family of homologous points associated with $(I)_n$ . With this definition we also have that j-th column of $M$ corresponds to homologous points of Frame $I_j$.

3. **Configuration**

   A configuration is a pair $((P)_n, \Omega)$ where $(P)_n = (P_1, \ldots, P_n)$ is a family of cameras and $\Omega = \{X_1, \ldots, X_m\}$, with $X_i \in \mathbb{R}P^3$, is a set of points in the scene.

4. **Explanation for families of homologous points**

   Established a tolerance $\varepsilon \in R^+$ , we define that a projective explanation for a family of homologous points $(x)_n = (x_1, \ldots, x_n)$ is a configuration $((P)_n, \Omega)$ such that $\forall i \in \{1, \ldots, n\}$, $\exists X_j \in \Omega$ which satisfies $d(P_i X_j, x_i) < \varepsilon$. In this case,

DOI: 10.1201/9781003206026-7

we say also that the configuration $((P)_n, \Omega)$ projectively explains the family of homologous points $(x)_n$.

5. **Explanation for matrices of homologous points**

   A projective explanation for a matrix of homologous points $M$ is a configuration that explains all families of homologous points of the lines of $M$. In this case, we also say that the configuration explains the projective matrix of homologous points $M$.

## 7.2 CALIBRATING IN PAIRS

It is not possible to immediately extend the pair calibration process of pairs of cameras, presented in Chapter 5, for a calibration of several cameras. The reason is the indetermination of the scale existing in each calibration pair by pair, as presented in Section 5.1.

For example, if we consider that we are in possession of a video $(I)_n$, and apply the calibration technique of Chapter 5, using the homologous points of the pairs $(I_1, I_2), (I_1, I_3), \ldots, (I_1, I_n)$, we will get, as an answer, pairs of cameras $(K[I|0], K[R_1|t_1]), (K[I|0], K[R_2|t_2]), \ldots, (K[I|0], K[R_{n-1}|t_{n-1}])$, where the directions and the senses of the vectors $t_1, t_2, \ldots, t_{n-1}$ , can be determined, but the values of $\|t_1\|, \|t_2\|, \ldots, \||t_{n-1}\|$ cannot.

The possibility of determining only the directions and sense of vectors $t_1, t_2, \ldots, t_{n-1}$, is a limitation of the calibration process carried out in pairs. the real scale indetermination, which is inherent to the problem of calibrating multiple cameras, is weaker. Although the values of $t_1, t_2, \ldots, t_{n-1}$ cannot be determined, the relationships $\frac{\|t_i\|}{\|t_j\|}$ can, that is, it is possible to find as an answer, a family of $n$ cameras of the form $(K[I|0], K[R_1|\lambda t_1], K[R_2|\lambda t_2], \ldots, K[R_{n-1}|\lambda t_{n-1}])$, where $\lambda \in \mathbb{R}^+$ is a factor that cannot be determined.

## 7.3 CALIBRATION IN THREE STEPS

We will now present an algorithm that finds a projective explanation $((P)_n, X_1, \ldots, X_m)$ for a matrix of homologous points $M$ associated with a video $(I)_n$. Although it was not a prominent focus, this algorithm appears as part of the calibration process described in [11].

The algorithm is formed by the following steps:

1. **Step 1**: Use the columns of $M$ corresponding to the homologous points of a pair of frames $I_i$ and $I_j$ to determine $P_i$ and $P_j$.

2. **Step 2**: Use the pair $P_i$ and $P_j$ and the matrix $M$ to determine the set $\{X_1, \ldots, X_m\}$.

3. **Step 3**: Use the set $X_1, \ldots, X_m$ and the matrix $M$ to determine the family of cameras $(P)_n$.

As presented in the previous chapters, the Step 1 can be solved by the Theorem 5.3, and the Steps 2 and 3 can be solved using Theorem 3.2, as explained in the Sections 5.8.2 and 4.1.2.

This three-step calibration process is interesting, as it avoids the use of sophisticated mathematical modeling based on trifocal tensors, an interesting property for our introductory book. A study on calibration made with trifocal tensors can be found in [13] and [9].

## 7.4 THREE-STEP CALIBRATION PROBLEMS

The naive implementation of three-step calibration shows bad results due to the following problems:

1. **Problem of Step 1** Gross errors may occur during the execution of the Step 1, as the fundamental matrix is estimated using a set of homologous points that can present gross errors, since we are considering that these are automatically determined by the KLT algorithm, which does not offer guarantees about its accuracy or correctness.

2. **Problem of Step 2** Gross errors may occur during the execution of Step 2 due to conditioning problems in the reconstruction process, as it is possible to that some reconstructed point of the scene is such that a great of disturbance its position in one direction causes a small change in the coordinates of the projections obtained by the cameras.

3. **Problem of Step 3** Step 3 does not impose the restriction given by the fact that the parameters intrinsic characteristics are known. Such parameters are used in step 1 when obtains the essential matrix from Equation 5.4.

4. **Limitations to small sequences** A very important problem of the three-step calibration is the fact that it is designed to solve the calibration to limited sequences of frames. It means that some extra processing is necessary in order to achieve a global optimization for the whole video.

We will show you how to solve the first three problems in order to make the calibration robust. For this, we will make use of the RANSAC algorithm. The problem of finding a good calibration for the whole video will be explained in Section 7.14.

## 7.5 MAKING THE CALIBRATION OF SMALL SEQUENCES ROBUST

We will present below how it is possible to use RANSAC to solve the problems in Steps 1, 2 and 3. We will recover this algorithm previously presented in Section 3.6.1 in order to make it simple to identificate the elements of it in the following sections:

"Given a model that requires a minimum of $n$ data points to instantiate its free parameters, and a set of data points $P$ such that the number of points in $P$ is greater than $n$ [$\sharp(P) \geqslant n$], randomly select a subset S1 of $n$ data points from $P$ and instantiate the model. Use the instantiated model $M1$ to determine the subset $S1*$ of points in

$P$ that are within some error tolerance of $M1$. The set $S1*$ is called the consensus set of $S1$.

If $\sharp(S1*)$ is greater than some threshold $t$, which is a function of the estimate of the number of gross errors in $P$, use $S1*$ to compute (possibly using least squares) a new model $M1*$.

If $\sharp(S1*)$ is less than $t$, randomly select a new subset $S2$ and repeat the above process. If, after some predetermined number of trials, no consensus set with t or more members has been found, either solve the model with the largest consensus set found, or terminate in failure."

The two columns of the matrix of homologous points $M$, corresponding to the homologous points used in the three-dimensional reconstruction made in Step 2, will be called base columns.

### 7.5.1 Solution to the Problem of Step 1

In this case, we can consider that the eight-point algorithm provides an way of getting the fundamental matrix, which corresponds to the model $M1$, from of a set formed by eight pairs of homologous points corresponding to $S1$, obtained in the base columns of $M$.

A tolerance criterion can be used to define the consensus set $S1*$, based on the objective function of the eight-point algorithm. More precisely, given a threshold $\eta_1 \in \mathbb{R}^+$ established empirically, we include in $S1*$ pair of homologous points $(x_i, x_j)$ of the base columns of $M$, if $|x'^T_i F x_j| < \eta_1$, where $F$ is the fundamental matrix taken using the set $S_1$ . The $M1*$ model is the fundamental matrix that can be obtained by applying the eight-point algorithm itself over the homologous points of $S1*$.

### 7.5.2 Solution to the Problem of Step 2

Let $Q$ be the set formed by the three-dimensional reconstructions of the pairs of homologous points of the base columns of $M$ that are part of the consensus set found during the application of RANSAC in the estimation of the fundamental obtained in Step 1.

In order to solve the conditioning problem in Step 2 we will use the RANSAC during the execution of Step 3. For that, the set $\Gamma$, formed by six pairs $(X, m)$, plays the role of the model $S1$, such that $X$ is an element of $Q$, and $m$ is the line of $M$ corresponding to the family of homologous points associated with $X$. The model $M1$ is a family of cameras $(P)_n$ obtained by the application of Step 3 using only the elements of $\Gamma$. The tolerance criterion used to define $S1*$ is based in the measure of the reprojection error. More precisely, given a threshold $\eta_2 \in \mathbb{R}^+$ chosen empirically, we insert in $S1*$ the pairs $(X', m')$ with $X' \in Q$ that satisfy, $\forall j \in \{1, \ldots, n\}, d(P_j X', m'_j) < \eta_2$ .

The $M1*$ model corresponds to a family of cameras $(P^*)_n$, estimated from the set $S1*$.

Thus, the set formed by the points $X'$ inserted in $S1*$ , and the family of cameras $(P^*)_n$ , define a projective explanation, satisfying the tolerance $\eta_2$ , for the matrix of homologous points $M'$, formed by lines of $M$.

### 7.5.3 Solution to the Problem in Step 3

Considering that the matrix of homologous points $M$ has $n$ columns, there are $(n^2 - n)/2$ possible choices for the base column pair. Therefore, we can try to solve the problem of Step 3, discarding the solution, if the intrinsic parameters of any of the cameras found is very different from the parameters that we are assuming. The three steps are repeated considering different choices of base columns until some good solution is found. More precisely, given an empirically chosen threshold $\eta_3 \in \mathbb{R}_+$, we refuse the family $(P^*)_n$ case $||K_j - K|| > \eta_3$, for some $j \in \{1, \ldots, n\}$, such that $K_j$ is a matrix of the intrinsic parameters obtained by the factorization of $P_j$ in the form $K_j[R_j|t_j]$, and $K$ is the matrix of intrinsic parameters that we are assuming as known. We can use, for example, the Frobenius norm in order to compare $K_j$ and $K$.

## 7.6 CHOICE OF BASE COLUMNS

As we have the possibility to choose $(n^2 - n)/2$ pairs of base columns for use in Steps 1 and 2, it makes sense to choose the one that provides the best result.

We can then define that the best result is the configuration that has not been discarded due to intrinsic parameter problems in Step 3 and that explains the largest number of lines of the matrix of homologous points $M$. A very efficient way to determine this pair was obtained using the following strategy:

1. We should not try to use base columns whose average distance from homologous points do not exceed a certain threshold.

2. If the number of pairs of homologous points obtained by RANSAC applied to Step 1 is less than the number of lines of $M$ explained by a configuration $C$, already calculated using another choice of base columns, we should abort the execution, as it is impossible for the $C$ configuration to be improved. This avoids we performed RANSAC in Step 2, which is the one with the highest computational cost.

3. We should first use columns of $M$ that are away as base columns, as they usually provide a better result than the nearby columns. This makes that the good results are determined before the bad ones, and with that we increased the effect of the previous item.

## 7.7 BUNDLE ADJUSTMENT

Now we will explain a process called Bundle Adjustment in the literature [13].

Let $((P)_n, X_1, \ldots, X_m)$ be a projective explanation for a matrix of homologous points $M$. We can define the reprojection error associated with this explanation as

$$\sum_{k=1}^{n} \sum_{i=1}^{m} d(P_k X_i, M_{ik})^2.$$

The smaller the reprojection error, the better the explanation is. Thereby, it makes sense to define the problem of finding an optimal projective explanation for a matrix of homologous points $M$. This problem can be solved using the Levenberg-Marquardt algorithm. In this case, the objective function is given by

$$g(x) = \frac{1}{2}||f(x) - x_0||^2,$$

in which $x_0 \in \mathbb{R}^{2mn}$ is a vector whose components are the coordinates of the projections of the $n$ points in the $m$ images obtained by the cameras, and the function $f : E^n \times \mathbb{R}^{3m} \to \mathbb{R}^{2mn}$ is defined by

$$(P_1, \ldots, P_n, X_1, \ldots, X_m) \mapsto (P_1X_1, \ldots, P_1X_m, \ldots, P_nX_1, \ldots, P_nX_m),$$

such that $E \subset \mathbb{R}^{12}$ is a space for representing virtual cameras. The set $E^n \times \mathbb{R}^{3m}$ is formed by representations of configurations of $n$ cameras and $m$ points.

## 7.8 REPRESENTATION OF A CONFIGURATION

A configuration of $m$ points and $n$ cameras can be represented by a vector of $\mathbb{R}^{12n+3m}$, in which $12n$ coordinates correspond to the elements of $n$ $3 \times 4$ matrices associated with the $n$ cameras, and $3m$ coordinates correspond to the coordinates of $m$ points of the three-dimensional scene. The problem with this representation, in our context, is that it does not impose the restriction characterized by the fact that the intrinsic parameters of the cameras are known. One way to impose this constraint is to use a vector of $\mathbb{R}^{6n+3m}$ as representation for a configuration. In this representation, the cameras have only six degrees of freedom, which correspond to extrinsic parameters. of these six degrees of freedom, three specify the rotation, which defines the orientation of the referential of the camera, and three specify the positioning of the projection center. This form of parameterization for the rotations is similar to that made in Section 4.5.2.

## 7.9 REFINEMENT CYCLES

One of the problems with three-step calibration is the possibility of some family of homologous points be discarded for presenting a reprojection error very high in some frame, due to the fact that the three-dimensional reconstruction performed by Step 2 only take into account a single pair of frames from the video. The solution adopted for this problem was to combine the calibration in three steps with a calibration made with the Levenberg-Marquardt algorithm.

Initially, a projective explanation $((P)_n, \Omega_1)$ is obtained by execution of the calibration in three steps using a threshold $\eta_2$ , defined in the Section 7.5.2, relatively high, chosen in such a way that a large number of families of homologous points are accepted even though some points with gross errors may contaminate the solution. This solution is then refined by an algorithm formed by cycles of four steps that are presented below, with the objective of selecting in a more judicious way the families of homologous points that should be considered in the estimation of the projective explanation.

1. Some iterations of the Levenbeg-Marquardt algorithm are performed, using as initial estimate the projective explanation $((P)_n, \Omega_1)$, determining another projective explanation $((P')_n, \Omega_2)$ of lower associated reprojection error.

2. Camera pairs of $(P')_n$ are used to determine a new reconstruction $\Omega_3$ for all homologous points of $M$. This process can be carried out by choosing different pairs of cameras to reconstruct each $\Omega_3$ point, so that each pair used is the one that minimizes the reprojection error associated to each point.

3. The points of $\Omega_3$ whose reprojection error in relation to the cameras $(P)_n$ are larger than a threshold $\eta_2'$ are discarded. We choose an $\eta_2'$ tighter than that $\eta_2$, that is, $\eta_2' < \eta_2$. This results in a new set of points $\Omega_4$ .

4. A new family of cameras $(P'')_n$ is estimated from the $\Omega_4$ and from the respective lines of the matrix of homologous points $M$. With this, we obtain a projective explanation $((P'')_n, \Omega_4)$ that can be used to feed a new refinement cycle.

The tolerance threshold for the reproduction error can be used for each cycle smaller and smaller, keeping in mind that as the solution gets more and more correct, we can be more and more rigorous. After executing a certain number of refinement cycles we can apply the Levenberg-Marquardt algorithm until its convergence, obtaining a projective explanation whose reprojection error associated with the families of selected homologous points is a local minimum.

## 7.10 EXAMPLE

The Figure 7.1 show frames of a video, with resolution $320 \times 240$, that have points tracked by the KLT algortithm. Let us analyse the effect of Refinement Cycles over the sequences of frames of it, in the cases which the KLT Tracker was used to track respectively: 50, 100 and 150 points. The results are presented in the graph of Figure 7.2.

The graph of Figure 7.2 present a curve which present the number of selected points in 5 different moments represented in the horizontal axis: A corresponds to the number of points selected in the first frame of the video by the KLT, B corresponds to the number of points tracked by the KLT over all the frames, C corresponds to the number of points selected by the RANSAC algorithm during the three steps

Figure 7.1 Sequence of frames tracked using the KLT tracker.

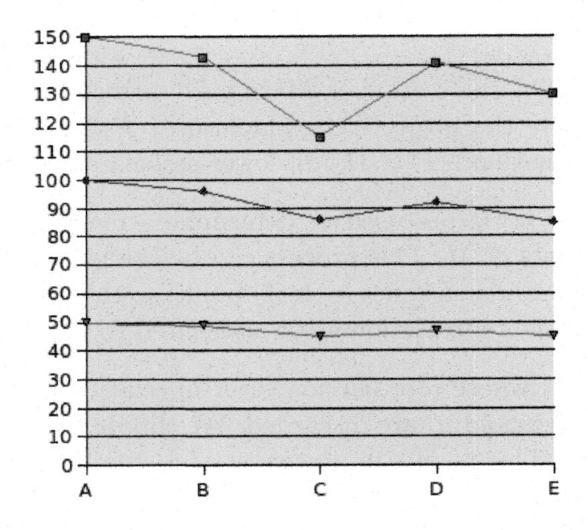


Figure 7.2 Number of points used in different steps of calibration in the fragments of the video of Figure 7.1.

calibration, D corresponds to the number of points selected by the first refinement cycle, E corresponds to the number of points selected by the second cycle of refinement.

The threshold used in the three steps calibration was 5 pixels. After that we applied a two refinement cycles, the first using a threshold of 3 pixels and the second using 2 pixels.

We can conclude from the graph of Figure 7.2 that many points discarded in the three steps calibration can be recovered by the application of refinement cycles even if the threshold used by them are narrow than the one used in the three steps calibration.

We also can conclude that when a larger number of features are tracked the refinemnt cycle is more important.

## 7.11 DECOMPOSITION OF THE VIDEO INTO FRAGMENTS

In a video $(I)_n$, it is possible that there are frames $I_a$ and $I_b$ that do not admit pairs of homologous points, in the case of no point of the scene is projected in both images. In addition, the KLT algorithms may not be able to correctly track points in long image sequences. As a consequence, it has to be said that it is not possible, in general, to define the matrix of homologous points for a complete video.

Taking advantage of the fact that the camera movement is continuous, it is possible to perform a decomposition of video $(I)_n$ into fragments, so that all fragments admit a matrix of homologous points. To be more precise, we are defining a fragment, of $k+1$ frames, of a video $(I_1, \ldots, I_n)$, as being a video of the form $(I_j, \ldots, I_{j+k})$, such that $\{j, j+1, \ldots, j+k\} \subset \{1, 2, \ldots, n\}$.

The fragments can be determined by a heuristic process. A possible solution consists on choose each fragment by comparing the frame $I_j$ with its successors until find a frame $I_{j+k}$ whose average distance of homologous pairs are higher than a threshold $\varepsilon \in \mathbb{R}_+$ , chosen experimentally. Thus obtaining a fragment of $k+1$ frames $(I_j, I_{j+1}, \ldots, I_{j+k})$.

In order to make it possible to join the fragments later, it is necessary to use a decomposition so that there is a superposition of a frame between each pair of adjacent fragments. That is, the video $(I)_k$ is decomposed into fragments of the form $(I_1, \ldots, I_{k_1})$, $(I_{k_1}, \ldots, I_{k_2})$, ...,$(I_{k_{n-2}}, \ldots, I_{k_{n-1}})$, $(I_{k_{n-1}}, \ldots, I_{k_n})$, in which each fragment is obtained as explained before.

It is possible that, when trying to determine the last fragment, the $\varepsilon$ threshold would not be satisfied, due to the reaching of the end of the video. In this case, these last frames must be discarderd, in order to avoid calibration problems by using a fragment related to a very small movement of camera.

## 7.12 JUNCTION OF FRAGMENTS

Let us consider that we found projective explanations for the matrices of homologous points of fragments of a video $(I)_n$. We will now show how to use use these explanations to determine a family of cameras $(P)_n$ corresponding to the cameras that were used to capture $(I)_n$ . It is necessary to take into account that each projective explanation was defined in its own referential, and in its own scale. So, let's divide the problem into two stages:

1. Alignment of fragments.
2. Compatibility of scales.

### 7.12.1 Alignment of Fragments

Given two configurations $E_1 = ((G)_r, \Omega)$ and $E_2 = ((Q)_s, \Psi)$, which explain the matrices of homologous points $M_1$ and $M_2$ , respectively associated with the consecutive fragments $F_1 = (I_k, I_{k+1}, \ldots, I_{k+r})$, and $F_2 = (I_{k+r}, I_{k+r+1}, \ldots, I_{k+r+s})$ of a video $(I)_n$. We want to determine a rigid movement that transforms $(Q)_s$ into a family $(Q')_s$, such that $G_r = Q'_1$ . We will say in this case that $(G)_r$ and $(Q)_s$ are aligned.

Let $Q_1 = K[R_1|t_1]$ and $G_r = K[R_2|t_2]$. We can determine the family $(Q)_s$ applying the following transformation to the elements of $(Q)_s$:

$$K[R|t] \mapsto K[(RR_1^T R_2)|RR_1^T(t_2 - t_1) + t].$$

We can use this transformation repeatedly to align all the family of cameras associated to each fragments of $(I)_n$.

### 7.12.2 Compatibility of Scales

The fact that two families of cameras $(G)_r$ and $(Q)_s$ associated to consecutive fragments are not aligned does not mean that they are ready to be concatenated in order to generate the family of cameras used to capture the two fragments. The reason is that generally $(G)_r$ and $(Q)_s$ are calibrated in different scales.

We can solve the problem of compatibility of scales by exploring the fact that given two projective explanations $E_1 = ((G)_r, \Omega)$ and $E_2 = ((Q)_s, \Psi)$ associated with consecutive fragments, there is usually a non-empty set $S \subset \Omega$ whose elements are points in the scene that also appear in $\Psi$. The scale factor $\lambda$ can be obtained as an answer to the following optimization problem

**Problem 7.12.1.** *Consider that $((G)_r, \Omega)$ and $((Q)_s, \Psi)$ are aligned fragments. Determine $\lambda \in \mathbb{R}_+$ such that by applying the transformation $K[R|t] \mapsto K[R|\lambda t]$ over all cameras in $(Q)_s$, a new fragment $((Q')_s, S)$ is obtained, which makes the reprojection error associated to points $S \subset \Omega$ that also appear in $\Psi$ be minimal.*

### 7.12.3 Robust Scale Compatibility

Solving the problem 7.12.1 is not simple, because as the coordinates of the elements of $S$ are estimated through a process of optimization associated to $((G)_r, \Omega)$, it is possible that some of the points of $S$ present gross reprojection errors when made by $(Q')_s$ cameras. This may occur if major changes in coordinates of points of $S$, in some direction, do not produce significant changes on the projections obtained by the cameras of $(G)_r$. In order to solve problem 7.12.1 in a robust way, we applied ideas present in the RANSAC algorithm, getting a solution in two steps:

1. Step 1: We find the set $\Lambda \subset \mathbb{R}_+$ formed by the values of $\lambda$ which maximizes the number of points of $S$ whose reprojection error relatve to the cameras $(Q')_s$ are below to a threshold $\xi \in \mathbb{R}_+$. These points of $S$ define the set $\Theta$;

2. Step 2: Solve the problem 7.12.1, modified by replacing set $S$ by its subset $\Theta$.

The Figure 7.3 illustrates the process of junction of fragments of the video presented in Figure 7.1.

In (a), the red curve shows the fraction of points reconstructed in the calibration of frames [0,54] tracked by the KLT in the fragment [54,95] such as the reprojection error in this frames are less than 5 pixels. The green curve shows the average error for the reprojection measured in pixels presented by the points represented by the red curve.

The graph presented in (b) has the same interpretation of the graph (a) considering the fragments [0,3] and [3,6].

By analyzing these graphs we can conclude that the problem of join finding the correct scale can be solved when we assume a sufficient number of frames in the fragment, but cannot be solved when we adopt a decomposition of very short fragments.

In the experiments represented by (a) and (b) 50 points have been tracked by KLT, such that 35 where selected by the refinement cycle in (a), and 49 have been selected by the cycles performed by (b). The little number of discarded points in (b) shows that lots of moving points have been wrongly accepted when the fragments are too shorts.

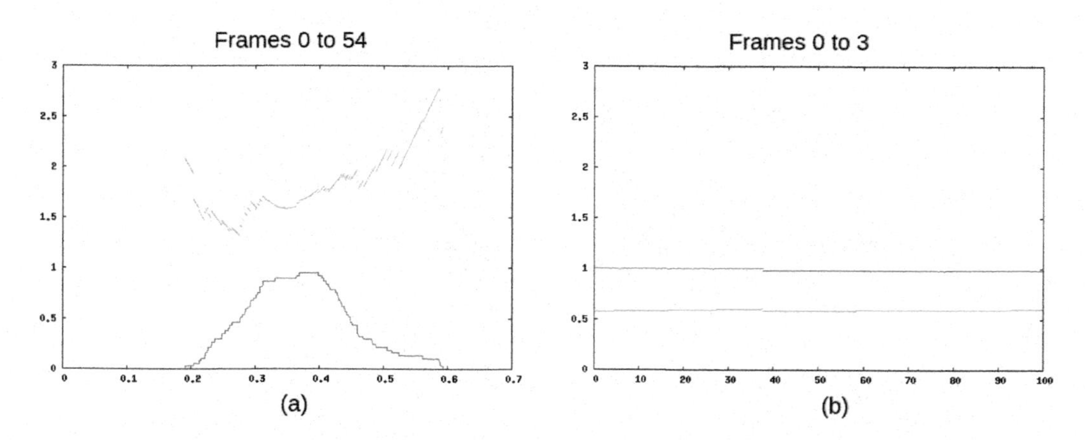


Figure 7.3 The red curve indicates the fraction of reconstructed points in the indicated fragment whose reprojection error in the fragments frame are inferior to 5 pixels. The green curve indicates the mean error in the reprojection. The information is parametrized by different choices of scales to the Problem 7.12.1. The result found in (a) is 0.368 and the result in (b) is ill-defined.

## 7.13 OFF-LINE AUGMENTED REALITY

After compatibilizing the scales of all frames, we are able to superimpose the frames of the videos with synthetic objects geometrically consist of them. The Figure 7.4 shows examples in which two cubes are rendered geometrically consistent of two sequences of frames

## 7.14 GLOBAL OPTIMIZATION BY RELAXATION

The algorithm presented in the previous section finds the junction of fragments of the video, each one calibrated in an optimal way. This means that we can accumulate errors in each junction since we are not finding a global solution to the entire video.

We can use the cameras found on this way in order to find a global optimization.

It can be done by successive steps that improve the quality of the global solution by a relaxation method that applies, alternately, two different kinds of operations:

Figure 7.4 Superimposing frames with synthetic objects geometrically consistent.

1. **Operation 1**: Making all the points fixed, use the Levenberg-Marquardt algorithm to optimize the parameters of one camera.

2. **Operation 2**: Making all the cameras fixed, optimize the 3D coordinates of a selected point using also the Levenberg-Marquardt algorithm.

After applying these operations, the reprojection error of all the points globally reduces. Thus we approximate our solution to the global optimal solution.

These two operations must be combined with an inlier/outlier classification in such a way that we choose the reconstruction that maximize the number of inliers, and if two situations have the same number of them, we choose the one that minimizes the reprojection error.

Before applying these two steps it is necessary to define the initial 3D coordinates associated to all tracked features. This procedure is necessary because each fragment of video defines a different set of 3D points related to the features. We can solve this problem comparing all reconstructions generated by each pair of cameras and choosing the reconstruction that maximize the number of inliers assuming some threshold for the reprojection error.

Our experiments have shown that the results achieved by this process is sufficiently to find the cameras with a precision good enough for making a good visual effect.

## 7.15 CODE MODULES

The code of this chapter is organized in the following modules:

1. Bundle Adjustment

2. RANSAC

3. Features List

4. Sequence of Frames

5. Relaxation

### 7.15.1 Bundle Adjustment API

```
gsl_vector *ba_ext_param_alloc( int n_cameras,
           int n_points );
```

This function creates a vector used for representing all cameras and all points in the Gnu Scientific Library. *n_cameras* is the number of cameras, and *n_points* is the number of points.

```
void ba_ext_get_rt( gsl_vector *r, gsl_vector *t,
           const gsl_vector *x, int n_points,
           int cam_index );
```

This function returns the extrinsic parameter $[r|t]$ given an output $x$ of the function *ba_ext_exec*. The parameter *cam_index* is the index of the camera, and *n_points* is the number of points encoded in $x$.

```
void ba_ext_get_camera( gsl_matrix *camera,
          const gsl_vector *x, int n_points, int cam_index,
          gsl_matrix *k );
```

This function returns a camera in the first parameter. $x$ receives a parameter that represents a configuration, *n_points* receives the number of points in $x$, *cam_index* receives the camera index and $k$ receives the matrix of intrinsic parameters.

```
void ba_ext_set_camera( gsl_vector *x, gsl_vector *r,
                 gsl_vector *t, int n_points,
                 int cam_index );
```

This function set the values of the camera of index *cam_index* in the vector of parameters $x$, received by *ba_ext_exec*. $r$ must contains the rotation matrix of the camera, and $t$ the translation.

```
void ba_get_proj( gsl_vector *prj, gsl_matrix *projs,
                  int cam_index, int point_index );
```

This function returns in *prj* the projection encoded in *projs* related to the point *point_index* and the camera *cam_index*.

```
void ba_set_proj( gsl_matrix *projs, gsl_vector *prj,
                  int cam_index, int point_index );
```

This function encodes in the matrix *projs* the projection *prj*, related to the point *point_index* and the camera *cam_index*.

```
void ba_get_point( gsl_vector *v, const gsl_vector *x,
                   int point_index );
```

This function returns in $v$ the point *point_index* encoded in the vector $x$.

```
void ba_set_point( gsl_vector *x, gsl_vector *v,
                   int point_index );
```

This function encodes in $x$ the point *point_indexed* with coordinates defined by $v$.

```
void ba_ext_exec( gsl_vector *xout, gsl_vector *xin,
                  gsl_matrix *projs, gsl_matrix *k );
```

This function executes the bundle adjustment algorithm considering the initial parameters encoded in the vector $xin$ and returns the parameters in $xout$. $projs$ is a $n \times 2$ matrix that contains all the 2D coordinates of all projections over $n$ frames, and $k$ is the intrinsic parameter of the camera.

```
int ba_ext_cost_func(const gsl_vector *x, void *params,
                     gsl_vector *f);
```

This function is used by the *ba_ext_exec* to return the optimization error in each iteration.

```
double ba_ext_reproj_error( const gsl_vector *x,
           gsl_matrix *projs, int cam_index,
           int point_index );
```

This function returns the reprojection error of the 3D point indexed by *point_index* related to the camera indexed by *cam_index*. The parameter $x$ constains all the information of the camera and the points of the scene.

```
void ba_set_lm_max_iterations( int n );
```

This function sets that $n$ is the maximum number of steps of the Levenberg-Marquardt that optimize the cameras and points.

```
void ba_optimize( gsl_multifit_fdfsolver *s );
```

This function executes the Bundle Adjustment algorithm. It is used by the function *ba_ext_exec*.

There are also the functions *ba_axis_angle_to_r* and *ba_r_to_axis_angle* that have already been defined in the Chapter 4.

### 7.15.2 Bundle Adjustment Code

```
// ba/ba.h

#ifndef BA_H
#define BA_H

#include "calib.h"
#include <gsl/gsl_multifit_nlin.h>

#define LM_MAX_ITERATIONS 50
#define LM_EPS 1e-2

#ifndef MAX
#define MAX(U,V)  (U>V?U:V)
#endif


typedef struct BundleAdjustData{
  gsl_matrix *projs;
  int n_cameras;
  int n_points;
} BundleAdjustData;


/* Extrinsic parameter Bundle Adjust */
gsl_vector *ba_ext_param_alloc( int n_cameras, int n_points );
void ba_ext_exec( gsl_vector *xout, gsl_vector *xin, gsl_matrix *projs,
                  gsl_matrix *k );
void ba_ext_get_rt( gsl_vector *r, gsl_vector *t, const gsl_vector *x,
                    int n_points, int cam_index );
void ba_ext_get_camera( gsl_matrix *camera, const gsl_vector *x,
                        int n_cams, int cam_index, gsl_matrix *k );
void ba_ext_set_camera( gsl_vector *x, gsl_vector *r, gsl_vector *t,
                        int n_points, int cam_index );
double ba_ext_reproj_error( const gsl_vector *x, gsl_matrix *projs,
                            int cam_index, int point_index );


/* ba.c */
void ba_get_proj( gsl_vector *prj, gsl_matrix *projs, int cam_index,
                  int point_index );
void ba_set_proj( gsl_matrix *projs, gsl_vector *prj, int cam_index,
                  int point_index );
void ba_get_point( gsl_vector *v, const gsl_vector *x, int point_index );
void ba_set_point( gsl_vector *x, gsl_vector *v, int point_index );
void ba_optimize( gsl_multifit_fdfsolver *s );
void ba_set_lm_max_iterations( int n );


/* Rotations */
void ba_axis_angle_to_r( gsl_matrix *r, gsl_vector *axis_angle );
void ba_r_to_axis_angle( gsl_vector *axis_angle, gsl_matrix *r );


#endif

// ba/ba_ext_coder.c


gsl_vector* ba_ext_param_alloc( int n_cameras, int n_points )
{
 return gsl_vector_alloc( 3*n_points + 6*n_cameras );
```

```
}


void ba_ext_get_rt( gsl_vector *r, gsl_vector *t, const gsl_vector *x,
                    int n_points, int cam_index )
{
 gsl_vector_set( r, 0, gsl_vector_get( x, 3*n_points + 6*cam_index ));
 gsl_vector_set( r, 1, gsl_vector_get( x, 3*n_points + 6*cam_index + 1 ));
 gsl_vector_set( r, 2, gsl_vector_get( x, 3*n_points + 6*cam_index + 2 ));
 gsl_vector_set( t, 0, gsl_vector_get( x, 3*n_points + 6*cam_index + 3));
 gsl_vector_set( t, 1, gsl_vector_get( x, 3*n_points + 6*cam_index + 4 ));
 gsl_vector_set( t, 2, gsl_vector_get( x, 3*n_points + 6*cam_index + 5 ));
}


void ba_ext_get_camera( gsl_matrix *camera, const gsl_vector *x,
                        int n_points, int cam_index, gsl_matrix *k )
{
 gsl_vector *h, *t;
 gsl_matrix *r;

 h = gsl_vector_alloc(3);
 t = gsl_vector_alloc(3);
 r = gsl_matrix_alloc(3,3);

 ba_ext_get_rt( h, t, x, n_points, cam_index );
 ba_axis_angle_to_r( r, h );
 calib_pmatrix_make( camera, k, r, t );

 gsl_vector_free(h);
 gsl_vector_free(t);
 gsl_matrix_free(r);
}


void ba_ext_set_camera( gsl_vector *x, gsl_vector *r, gsl_vector *t,
                        int n_points, int cam_index )
{
 gsl_vector_set( x, 3*n_points + 6*cam_index, gsl_vector_get( r, 0 ));
 gsl_vector_set( x, 3*n_points + 6*cam_index + 1, gsl_vector_get( r, 1 ));
 gsl_vector_set( x, 3*n_points + 6*cam_index + 2, gsl_vector_get( r, 2 ));
 gsl_vector_set( x, 3*n_points + 6*cam_index + 3, gsl_vector_get( t, 0 ));
 gsl_vector_set( x, 3*n_points + 6*cam_index + 4, gsl_vector_get( t, 1 ));
 gsl_vector_set( x, 3*n_points + 6*cam_index + 5, gsl_vector_get( t, 2 ));
}


// ba/ba_ext.c


static int ba_ext_cost_func(const gsl_vector *x, void *params,
                            gsl_vector *f);
static int ba_ext_diff_func(const gsl_vector *x, void *params,
                            gsl_matrix *J);

static gsl_matrix *kmatrix;


void ba_ext_exec( gsl_vector *xout, gsl_vector *x,
                  gsl_matrix *projs, gsl_matrix *k )
{
 const gsl_multifit_fdfsolver_type *t = gsl_multifit_fdfsolver_lmsder;
 gsl_multifit_fdfsolver *s;
 gsl_multifit_function_fdf f;
```

```
 BundleAdjustData d;

 d.projs = projs;
 d.n_points = projs->size1;
 d.n_cameras = projs->size2/2;

 f.f = &ba_ext_cost_func;
 f.df = NULL;
 kmatrix = k;

 f.p = 3*d.n_points + 6*d.n_cameras ;
 f.n = d.n_cameras * d.n_points;
 f.params = &d;

 s = gsl_multifit_fdfsolver_alloc(t, f.n, f.p);
 gsl_multifit_fdfsolver_set(s, &f, x);

 ba_optimize(s);

 gsl_vector_memcpy( xout, s->x );
 gsl_multifit_fdfsolver_free(s);
}


int ba_ext_cost_func(const gsl_vector *x, void *params, gsl_vector *f)
{
 int i,j;
 BundleAdjustData *data;

 data = (BundleAdjustData*)params;
 for( i = 0; i < data->n_cameras; i++ )
   for( j = 0; j < data->n_points; j++ ){
      gsl_vector_set( f, i*data->n_points + j,
                      ba_ext_reproj_error( x, data->projs, i, j ) );
   }

 return GSL_SUCCESS;
}


/* error between estimated projection and real image projection*/
double ba_ext_reproj_error( const gsl_vector *x, gsl_matrix *projs,
                            int cam, int point_index )
{
 double error;
 gsl_vector *prj, *v, *v_prj;
 gsl_matrix *p;

 v = gsl_vector_alloc(3);
 v_prj = gsl_vector_alloc(2);
 prj = gsl_vector_alloc(2);
 p = gsl_matrix_alloc(3,4);

 ba_get_proj( prj, projs, cam, point_index );
 ba_ext_get_camera( p, x, projs->size1, cam, kmatrix );
 ba_get_point( v , x, point_index );
 calib_apply_P( p, v, v_prj );
 gsl_vector_sub( prj, v_prj );
 error = SQR( gsl_blas_dnrm2( prj ) );

 gsl_vector_free(v);
 gsl_vector_free(v_prj);
 gsl_vector_free(prj);
 gsl_matrix_free(p);
```

```
 return error;
}

// ba/ba.c


static int lm_max_iterations = LM_MAX_ITERATIONS;


void ba_get_proj( gsl_vector *prj, gsl_matrix *projs, int cam_index,
                  int point_index )
{
 gsl_vector_set( prj, 0, gsl_matrix_get( projs, point_index, 2*cam_index ) );
 gsl_vector_set( prj, 1, gsl_matrix_get( projs, point_index, 2*cam_index + 1 ) );
}


void ba_set_proj( gsl_matrix *projs, gsl_vector *prj, int cam_index,
                  int point_index )
{
 gsl_matrix_set( projs, point_index, 2*cam_index, gsl_vector_get( prj, 0 ) );
 gsl_matrix_set( projs, point_index, 2*cam_index + 1, gsl_vector_get( prj, 1 ));
}


void ba_get_point( gsl_vector *v, const gsl_vector *x, int point_index )
{
 gsl_vector_set( v, 0, gsl_vector_get( x, 3*point_index ));
 gsl_vector_set( v, 1, gsl_vector_get( x, 3*point_index + 1));
 gsl_vector_set( v, 2, gsl_vector_get( x, 3*point_index + 2));
}


void ba_set_point( gsl_vector *x, gsl_vector *v, int point_index )
{
 gsl_vector_set( x, 3*point_index , gsl_vector_get(v, 0) );
 gsl_vector_set( x, 3*point_index + 1 , gsl_vector_get(v,1) );
 gsl_vector_set( x, 3*point_index + 2 , gsl_vector_get(v,2) );
}


void ba_set_lm_max_iterations( int n )
{
 lm_max_iterations = n;
}


void ba_optimize( gsl_multifit_fdfsolver *s )
{
 int status, iter = 0;

 do{
    iter++;
    status = gsl_multifit_fdfsolver_iterate(s);
    if(status)
       break;
    status = gsl_multifit_test_delta(s->dx, s->x, LM_EPS, LM_EPS);
    printf( "%i Error = %f\n", iter, gsl_blas_dnrm2(s->f) );
 }
 while( (status == GSL_CONTINUE) && (iter <  lm_max_iterations ) );
}
```

### 7.15.3 RANSAC API

```
int ransac_fmatrix_make( int *inliers, gsl_matrix *f,
        gsl_matrix *a, gsl_matrix *b,  int n_iters,
        double tol );
```

This function implements the algorithm described in Section 7.5.1. It implements a robust estimation of a Fundamental Matrix $f$, givem a pair of homologous corrupted by outliers encoded in the lines of the $2 \times n$ matrices $a$ and $b$. The RANSAC algorithm execute $n\_iters$ iterations and it accepts points that makes $|x_i F x_j| < tol$. The $inliers$ array contain the information of the index of the inliers and outliers points. A 0 value means an outlier and the 1 value means a inlier point.

```
int ransac_reconstruct( int *inliers,  gsl_matrix *points,
                        gsl_matrix *projs1,
                        gsl_matrix *projs2, gsl_matrix *k,
                        int n_iters, double tol );
```

This function implements the algorithm described in Section 7.5.2. It implements a robust 3D reconstrution of points that are encoded as the output in the $n \times 3$ matrix $points$. The considered 2D homologous points are encoded in the $2 \times n$ matrices $projs1$ and $projs2$. The intrinsic parameter considered is $k$. The number of iterations of the ransac is $n\_iters$ and the tolerance for the reprojection error is $tol$. The $inliers$ works in the same way as the correspondent in the function $ransac\_fmatrix\_make$.

```
int ransac_singlecam( int *inliers, gsl_matrix *p,
   gsl_matrix *q, gsl_matrix *x, int n_iters, double tol );
```

This function can be used to find, in a robust way, a $3 \times 4$ camera matrix $p$, given a set of 3D points encoded in the $q$ in the form of a $n \times 3$ matrix, $x$ encodes the respective 2D projections in the form of a $n \times 2$ matrix. $n\_iters$ defines the number of iterations of the RANSAC algorithm, $tol$ defines the reprojection tolerance, and inliers is an array that works in the same way of the previous function.

```
void get_random_choose( int *samples, int nsamples, int n );
```

This is an auxiliary function used by the RANSAC algorithm to choose a set of elements to be used. $n$ encodes the number of elements of the whole set, *nsamples* defines the size of the subset and *samples* returns the selected index of the element.

```
void get_inliers( gsl_matrix *dst, gsl_matrix *src,
                  int *inliers );
```

This is an auxiliary function. It selects the inliers rows of the matrix *src* represented in the *inliers*, in such a way that 1 represents an inlier and 0 represents an outlier. The selected elements are copied in the *dst* matrix.

```
void get_samples( gsl_matrix *aux, gsl_matrix *m,
                  int *samples, int n );
```

This is an auxiliary function. It copies the *n* rows of the matrix *m* encoded in the *samples* array. The result is outputted in the rows of the *aux* matrix.

### 7.15.4 RANSAC Code

```
// ransac/ransac.h

#ifndef RANSAC_H
#define RANSAC_H

#include <defs.h>
#include <string.h>
#include "random.h"
#include "calib.h"

#define INLIER 1
#define OUTLIER 0

int ransac_fmatrix_make(  int *inliers, gsl_matrix *f, gsl_matrix *a,
                          gsl_matrix *b,  int n_iters, double tol );
int ransac_reconstruct(  int *inliers,  gsl_matrix *points, gsl_matrix *projs1,
                         gsl_matrix *projs2, gsl_matrix *k,  int n_iters,
                         double tol );
int ransac_singlecam( int *inliers, gsl_matrix *p, gsl_matrix *q,
                      gsl_matrix *x, int n_iters, double tol  );
int ransac_extrinsic_singlecam( int *inliers, gsl_matrix *p, gsl_matrix *q,
                                gsl_matrix *x, int n_iters, double tol,
                                gsl_matrix *k );

/* misc */
void get_random_choose( int *samples, int nsamples, int n );
void get_inliers( gsl_matrix *dst, gsl_matrix *src, int *inliers );
void get_samples( gsl_matrix *aux, gsl_matrix *m, int *samples, int n );

#endif

// ransac/fmatrix.c

static int consensus( int *inliers, gsl_matrix *f, gsl_matrix *a,
                      gsl_matrix *b, double tol );
static void get_homologus( gsl_matrix *x1, gsl_matrix *x2,
                           gsl_matrix *a, gsl_matrix *b, int i );


int ransac_fmatrix_make( int *inliers, gsl_matrix *f, gsl_matrix *a,
                         gsl_matrix *b, int n_iters, double tol )
{
 int n, j, samples[9], *aux_inliers;
 int c, best_consensus = 0;
 gsl_matrix *aux1, *aux2, *a_inliers, *b_inliers;
```

```
 n = a->size1;
 aux1 = gsl_matrix_alloc(9,2);
 aux2 = gsl_matrix_alloc(9,2);
 aux_inliers = (int*)malloc( n*sizeof(int) );

 for( j=0; j < n_iters; j++ ){
   get_random_choose( samples, 9, n-1 );
   get_samples( aux1, a, samples, 9 );
   get_samples( aux2, b, samples, 9 );
   calib_fmatrix_dlt( f, aux1, aux2 );
   if((c = consensus( aux_inliers, f, a, b, tol )) > best_consensus ) {
      best_consensus = c;
      memcpy( inliers, aux_inliers, n*sizeof(int));
   }
 }

 if( best_consensus > 8 ){
   a_inliers = gsl_matrix_alloc( best_consensus, 2 );
   b_inliers = gsl_matrix_alloc( best_consensus, 2 );
   get_inliers( a_inliers, a, inliers );
   get_inliers( b_inliers, b, inliers );
   calib_fmatrix_dlt( f, a_inliers, b_inliers );
   gsl_matrix_free( a_inliers );
   gsl_matrix_free( b_inliers );
 }

 gsl_matrix_free( aux1 );
 gsl_matrix_free( aux2 );
 free( aux_inliers );
 return best_consensus;
}


int consensus( int *inliers, gsl_matrix *f, gsl_matrix *a,
               gsl_matrix *b, double tol )
{
  int i, ninliers = 0;
  gsl_matrix *x1, *x2, *u, *error;

  x1 = gsl_matrix_alloc(3,1);
  x2 = gsl_matrix_alloc(1,3);
  u = gsl_matrix_alloc(3,1);
  error = gsl_matrix_alloc(1,1);

 for( i = 0; i < a->size1; i++ ){
     get_homologus( x1, x2, a, b, i  );
     gsl_linalg_matmult( f, x1, u );
     gsl_linalg_matmult( x2, u, error );
     if(  fabs( gsl_matrix_get( error, 0, 0 ))  < tol ){
          ninliers++;
          inliers[i] = INLIER;
     }
     else
          inliers[i] = OUTLIER;
 }

 gsl_matrix_free( x1 );
 gsl_matrix_free( x2 );
 gsl_matrix_free( u );
 gsl_matrix_free( error );
 return ninliers;
}
```

```
void get_homologus( gsl_matrix *x1, gsl_matrix *x2,
                    gsl_matrix *a, gsl_matrix *b, int i )
{
 gsl_matrix_set( x1, 0, 0, gsl_matrix_get(  a, i, 0 ) );
 gsl_matrix_set( x1, 1, 0, gsl_matrix_get(  a, i, 1 ) );
 gsl_matrix_set( x1, 2, 0, 1. );

 gsl_matrix_set( x2, 0, 0, gsl_matrix_get(  b, i, 0 ) );
 gsl_matrix_set( x2, 0, 1, gsl_matrix_get(  b, i, 1 ) );
 gsl_matrix_set( x2, 0, 2, 1. );
}


// ransac/reconstruct.c

int ransac_reconstruct(  int *inliers,  gsl_matrix *points,
                         gsl_matrix *projs1, gsl_matrix *projs2, gsl_matrix *k,
                         int n_iters, double tol )
{
 int i, ninliers;
 gsl_matrix *f, *e, *p, *pid, *id;
 gsl_vector *zero, *x1, *x2, *x;

 f = gsl_matrix_alloc( 3, 3 );
 e = gsl_matrix_alloc( 3, 3 );
 p = gsl_matrix_alloc( 3, 4 );
 pid = gsl_matrix_alloc( 3, 4 );
 id = gsl_matrix_alloc( 3, 3 );
 zero = gsl_vector_alloc( 3 );
 x1 = gsl_vector_alloc( 2 );
 x2 = gsl_vector_alloc( 2 );
 x = gsl_vector_alloc( 3 );

 gsl_matrix_set_identity( id );
 gsl_vector_set_zero( zero );

 ninliers = ransac_fmatrix_make(  inliers, f, projs1 , projs2, n_iters, tol );
 if( ninliers > 8 ){
   calib_ematrix_make( e, f, k );
   for( i=0; inliers[i] == OUTLIER;  i++);
   gsl_matrix_get_row( x1, projs1, i );
   gsl_matrix_get_row( x2, projs2, i );
   calib_ematrix_get_P( k, e, x1, x2, p );
   calib_pmatrix_make( pid, k, id, zero );

   for( i = 0; i < projs1->size1; i++ ) {
      gsl_matrix_get_row( x1, projs1, i );
      gsl_matrix_get_row( x2, projs2, i );
      calib_get_3dpoint( pid, p, x1, x2, x  );
      gsl_matrix_set( points, i, 0, gsl_vector_get( x, 0 ) );
      gsl_matrix_set( points, i, 1, gsl_vector_get( x, 1 ) );
      gsl_matrix_set( points, i, 2, gsl_vector_get( x, 2 ) );
   }
 }

 gsl_matrix_free( f );
 gsl_matrix_free( e );
 gsl_matrix_free( p );
 gsl_matrix_free( pid );
 gsl_matrix_free( id );
 gsl_vector_free( zero );
 gsl_vector_free( x1 );
 gsl_vector_free( x2 );
 gsl_vector_free( x );
```

```
 return  ninliers;
}


// ransac/singlecam.c

static double reproj_error( gsl_matrix *proj1,  gsl_matrix *proj2, int index );
static void  apply_P( gsl_matrix *p, gsl_matrix *points, gsl_matrix *reprojs );


int ransac_singlecam( int *inliers, gsl_matrix *p, gsl_matrix *q,
                      gsl_matrix *x, int n_iters, double tol  )
{
 int n, j, samples[6], *aux_inliers;
 int c, best_consensus = 0;
 gsl_matrix *qaux, *xaux, *q_inliers, *x_inliers;

 if( x->size1 > 5 )
   n = x->size1;
 else
   return 0;


 qaux = gsl_matrix_alloc(6,3);
 xaux = gsl_matrix_alloc(6,2);
 aux_inliers = (int*)malloc( n*sizeof(int) );


 for( j=0; j < n_iters; j++ ){
   get_random_choose( samples, 6, n-1 );
   get_samples( xaux, x, samples, 6 );
   get_samples( qaux, q, samples, 6 );

   calib_singlecam_dlt( p, qaux, xaux );

   if((c = consensus( aux_inliers, p, q, x, tol )) > best_consensus ) {
      best_consensus = c;
      memcpy( inliers, aux_inliers, n*sizeof(int));
   }

 }

 if( best_consensus > 5 ){
   q_inliers = gsl_matrix_alloc( best_consensus, 3 );
   x_inliers = gsl_matrix_alloc( best_consensus, 2 );
   get_inliers( q_inliers, q, inliers );
   get_inliers( x_inliers, x, inliers );
   calib_singlecam_dlt( p, q_inliers, x_inliers );
   gsl_matrix_free( q_inliers );
   gsl_matrix_free( x_inliers );
 }

 gsl_matrix_free( qaux );
 gsl_matrix_free( xaux );
 free( aux_inliers );
 return best_consensus;
}


int consensus( int *inliers, gsl_matrix *p, gsl_matrix *q,
               gsl_matrix *x, double tol )
{
  int i, ninliers = 0;
  gsl_matrix *x_proj;
```

```
   x_proj = gsl_matrix_alloc( x->size1, 2 );
   apply_P( x_proj, p, q );
   for( i = 0; i < x->size1; i++ ){
      if(  reproj_error( x, x_proj, i )   < tol ){
           ninliers++;
           inliers[i] = INLIER;
      }
      else
           inliers[i] = OUTLIER;
  }

  gsl_matrix_free( x_proj );
  return ninliers;
}


void  apply_P( gsl_matrix *reprojs, gsl_matrix *p, gsl_matrix *points )
{
 int i;
 gsl_vector *x, *x_proj;

 x =  gsl_vector_alloc( 3 );
 x_proj = gsl_vector_alloc( 2 );

 for( i = 0; i < points->size1; i++ ){
   gsl_matrix_get_row( x, points, i );
   calib_apply_P( p,  x,  x_proj );
   gsl_matrix_set_row( reprojs, i, x_proj );
 }

 gsl_vector_free( x );
 gsl_vector_free( x_proj );
}


double reproj_error( gsl_matrix *proj1,  gsl_matrix *proj2, int index )
{
 int i;
 double d = 0;
 gsl_vector *v1, *v2;

 v1 = gsl_vector_alloc( 2 );
 v2 = gsl_vector_alloc( 2 );
 gsl_matrix_get_row( v1, proj1, index );
 gsl_matrix_get_row( v2, proj2, index );
 gsl_vector_sub( v1, v2 );
 d = gsl_blas_dnrm2( v1 );

 gsl_vector_free( v1 );
 gsl_vector_free( v2 );
 return d;
}


// ransac/misc.c

void get_random_choose( int *samples, int nsamples, int n )
{
 int i, r, k = 0;
 while( k< nsamples ){
   r = (int)( n * uniform_random() + .5 );
   for( i = 0; (samples[i] != r) && (i<k); i++ );
   if( i==k ){
```

```
      samples[k] = r;
      k++;
    }
  }
}


void get_inliers( gsl_matrix *dst, gsl_matrix *src, int *inliers )
{
 int i, j, k = 0;

 for( i = 0; i < src->size1; i++ ){
    if( inliers[i] == INLIER ){
       for( j = 0; j < src->size2; j++ )
            gsl_matrix_set( dst, k, j, gsl_matrix_get( src, i, j ) );
       k++;
    }
 }
}


void get_samples( gsl_matrix *aux, gsl_matrix *m, int *samples, int n )
{
 int i, j;

 for( j=0; j < m->size2; j++ )
     for( i = 0; i < n; i++ )
        gsl_matrix_set( aux, i, j, gsl_matrix_get( m, samples[i], j ) );
}

// random/random.h

#ifndef RANDOM_H
#define RANDOM_H


#include <stdlib.h>
#include <time.h>
#include <stdio.h>

#define RND() (uniform_random())

void init_timer_random( void );
double uniform_random( void );

#endif

// random/random.c

#include "random.h"


double uniform_random(void)
{
 return (double)rand()/(double) RAND_MAX;
}


void init_timer_random( void )
{
  srand(time(NULL));
}
```

### 7.15.5 Features List API

```
Features *features_alloc( int nfeatures, int frame,
                          int *status );
```

This function allocates a structure of the type *Features*. The first parameter specifies the number of features in the structure. The second one specifies the *frame* that these features are related. The *status* array encodes which features are succefully tracked. If an element of the *status* array is 0 it means that the feature has been tracked, and if it is $-1$ it means that it has not been tracked.

```
Features *features_clone( Features *f );
```

This function returns a copy of $f$.

```
FeaturesList *features_list_alloc( void );
```

This function creates a void features list.

```
void features_dispose( Features *f );
```

This function deallocates $f$.

```
void features_list_dispose( FeaturesList *fl );
```

This function erases all the features list and deallocates $fl$.

```
void features_list_insert( FeaturesList *fl, Features *f );
```

This function inserts $f$ in the tail of the FeaturesList $fl$.

```
FeaturesList *features_list_read( FILE *fin );
```

This function processes the output of the program presented in the Section 6.7 and returns a FeaturesList that correspond to it.

```
void features_list_write( FILE *fout, FeaturesList *fl );
```

This function encodes in the file $fout$ the features list $fl$.

```
void features_reduce_status( Features *f1, Features *f2 );
```

This function compares the features of $f1$ and $f2$ and changes the status of features of both. More precisely, it makes the status of both equals to $-1$ (not tracked) if the status of one of them is already $-1$.

```
Real features_distancy( Features *f1, Features *f2 );
```

This function returns the average distance of the features tracked both by $f1$ and $f2$.

```
void features_list_begin_frag( FeaturesList *fl );
```

This function makes a static variable $fcurrent$ equals to $fl \rightarrow head$.

```
FeaturesList *features_list_get_next_frag( FeaturesList *fl,
                                 int nframes, double distancy );
```

This function returns a *FeaturesList* that contains the features of a sequence in which the average distance of them are higher than *distance*. If the number of frames in the sequence is higher than $nframes$ it cuts the FeaturesList even if the minimun average distance have not been achieved. After running this function the $fcurrent$ static variable is actualized to the first feature in the list not inserted in the return.

```
int features_list_end_frag( void );
```

This function returns TRUE if $fcurrent$ is $NULL$ and $FALSE$ otherwise.

### 7.15.6 Features List Code

```
// features/featureslist.h

#ifndef FEATURES_H
#define FEATURES_H

#include "geom.h"

#define FEATURE_TRACKED    0
#define FEATURE_VOID       -1

typedef struct Features{
   struct Features *next, *prev;
   int nfeatures;
   int frame;
   int *status;
   Vector3 *p;
}Features;

typedef struct FeaturesList{
```

```
    int nframes;
    Features *head, *tail;
}FeaturesList;


/* constructors & destructors */

Features *features_alloc( int nfeatures, int frame, int *status );
Features *features_clone( Features *f );
FeaturesList *features_list_alloc( void );

void features_dispose( Features *f );
void features_list_dispose( FeaturesList *fl );

/* insert */

void features_list_insert( FeaturesList *fl, Features *f );

/* persistence */

FeaturesList *features_list_read( FILE *fin );
void features_list_write( FILE *fout, FeaturesList *fl );

/* operations */

void features_reduce_status( Features *f1, Features *f2 );
Real features_distancy( Features *f1, Features *f2 );
FeaturesList *features_list_sel_sequence( Real d, int ncommon, Features *base );

/* fragment */

void features_list_begin_frag( FeaturesList *fl );
FeaturesList *features_list_get_next_frag( FeaturesList *fl, int nframes,
          double distancy );
int features_list_end_frag( void );

#endif

// features/alloc.c

#include "featureslist.h"


/* Constructors */

Features *features_alloc( int nfeatures, int frame, int *status )
{
 int i;
 Features *f;

 f = NEWSTRUCT( Features );
 f->nfeatures = nfeatures;
 f->next = f->prev =  NULL;
 f->frame = frame;
 f->p = NEWTARRAY( nfeatures, Vector3 );
 f->status = NEWTARRAY( nfeatures, int );
 if( status != NULL )
   for( i=0; i < nfeatures; i++ )
      f->status[i] = status[i];
 return f;
}


Features *features_clone( Features *f )
```

```
{
 int i;
 Features *fclone;

 fclone = NEWSTRUCT( Features );
 fclone->nfeatures = f->nfeatures;
 fclone->next = fclone->prev = NULL;
 fclone->frame = f->frame;
 fclone->p = NEWTARRAY( f->nfeatures, Vector3 );
 fclone->status = NEWTARRAY( f->nfeatures, int );
 for( i = 0; i < f->nfeatures; i++ ){
    fclone->p[i] = f->p[i];
    fclone->status[i] = f->status[i];
 }
 return fclone;
}


FeaturesList *features_list_alloc( void )
{
 FeaturesList *fl;

 fl = NEWSTRUCT( FeaturesList );
 fl->head = fl->tail = NULL;
 return fl;
}


/* Destructors */

void features_dispose( Features *f )
{
 efree( f->p );
 efree( f->status );
 efree( f );
}


void features_list_dispose( FeaturesList *fl )
{
 Features *aux1, *aux2;

 aux1 = fl->head;
 while( aux1 != NULL ){
   aux2 = aux1;
   aux1 = aux1->next;
   features_dispose(aux2);
 }
 efree( fl );
}

// features/insert.c

#include "featureslist.h"

void features_list_insert( FeaturesList *fl, Features *f )
{
 fl->nframes++;
 if( fl->head != NULL ){
   fl->tail->next = f;
   f->prev = fl->tail;
   fl->tail = f;
 }
 else
```

```
   fl->head = fl->tail = f;
}

// features/persist.c

#include "featureslist.h"


FeaturesList *features_list_read( FILE *fin )
{
 int i, nfeatures, frame;
 Real x, y;
 Features *f;
 FeaturesList *fl;

 fscanf( fin, "Features per frame = %i\n", &nfeatures );
 fl = features_list_alloc();

 while( !feof( fin ) ){
   fscanf( fin, "Frame %i\n", &frame );
   f = features_alloc( nfeatures, frame, NULL );
   for( i = 0; i < nfeatures; i++ )
      fscanf( fin, "%lf %lf %i\n", &(f->p[i].x), &(f->p[i].y), &f->status[i] );
   features_list_insert( fl, f );
 }
 return fl;
}

void features_list_write( FILE *fout, FeaturesList *fl )
{
 int i, nfeatures;
 Features *f;

  if( fl->head != NULL ){
    nfeatures = fl->head->nfeatures;
    fprintf( fout, "Features per frame = %i\n", nfeatures  );
    for( f = fl->head; f != NULL; f = f->next ){
        fprintf( fout, "Frame %i\n", f->frame );
        for( i=0; i < nfeatures; i++ )
          fprintf( fout, "%lf %lf %i\n", f->p[i].x, f->p[i].y, f->status[i] );
    }
 }
}

// features/opers.c

#include "featureslist.h"

#define BOTH_TRACKED ((f1->status[i] == FEATURE_TRACKED) && (f2->status[i]
          == FEATURE_TRACKED))


int features_ncommon( Features *f1, Features *f2 )
{
 int i;
 int ncommon = 0;

 for( i=0; i < f1->nfeatures; i++ )
   if( BOTH_TRACKED )
     ncommon++;
 return ncommon;
}
```

```
Real features_distancy( Features *f1, Features *f2 )
{
 int i, ncommon = 0;
 Real d = 0;

 for( i=0; i < f1->nfeatures; i++ ){
    if( BOTH_TRACKED ){
       d += v3_norm( v3_sub( f1->p[i], f2->p[i] ) );
       ncommon++;
    }
 }

 return d/ncommon;
}


void features_reduce_status( Features *f1, Features *f2 )
{
 int i;
 int ncommon = 0;

 for( i=0; i < f1->nfeatures; i++ )
   if( !BOTH_TRACKED ){
     f1->status[i] = FEATURE_VOID;
     f2->status[i] = FEATURE_VOID;
   }
}

// features/frag.c

#include "featureslist.h"

static Features *fcurrent;


void features_list_begin_frag( FeaturesList *fl )
{
 fcurrent = fl->head;
}


FeaturesList *features_list_get_next_frag( FeaturesList *fl, int nframes,
          double distancy )
{
 Features *faux, *fbase;
 FeaturesList *flout;

 flout = features_list_alloc();
 fbase = fcurrent;
 while( (fcurrent != NULL) && ((nframes >= 0) ||
    features_distancy(fbase,faux) < distancy) ){
      faux = features_clone( fcurrent );
      features_reduce_status( fbase, faux );
      features_list_insert( flout, faux );
      fcurrent = fcurrent->next;
      nframes--;
 }
 if( fcurrent != NULL  )
   fcurrent = fcurrent->prev;
 return flout;
}


int features_list_end_frag( void )
```

```
{
 if( fcurrent == NULL  )
   return TRUE;
 else
   return FALSE;
}
```

### 7.15.7 Sequence of Frames API

```
SubSeqNode *subseq_node_alloc( CalibSequenceContext *cc,
                               FeaturesList *fl );
```

This function allocates a structure *SubSecNode*, given a structure *CalibSequenceContext cc* and a structure *FeaturesList fl*.

```
void camera_fields_alloc( Camera *cam );
```

This function allocates the parameters $k$, $r$ and $t$ of a data structure *Camera cam*.

```
void camera_fields_dispose( Camera *cam );
```

This function deallocates the parameters $k$, $r$ and $t$ of a data structure *Camera cam*.

```
void subseq_node_dispose( SubSeqNode *sn );
```

This function deallocates a *structureSubSecNode sn*.

```
CalibSequenceContext *calib_context_alloc( FeaturesList
          *fl );
```

This function creates a structure *CalibSequenceContext* based on the *FeaturesList fl*.

```
void seq_extrinsic_calibrate( CalibSequenceContext *cc,
                              gsl_matrix *k  );
```

This function executes all the processing described from Section 7.3 to 7.12.3. After that, a first calibration for the whole cameras is estimated, and the result is stored in the structure *cc*. This is the function used in the main code program, described in the Section 7.16.

```
void extract_rt( gsl_matrix *r, gsl_vector *t,
                 gsl_matrix *ref );
```

This function extracts the matrix of rotation $r$ and the translation vector $t$, given a $3 \times 4$ matrix $[r|t]$ encoded in the matrix $ref$.

```
void calib_context_dispose( CalibSequenceContext *cc );
```

This function destroys the structure $CalibSequenceContext$ $cc$.

```
void subseq_extrinsic_calibrate( SubSeqNode *s,
          gsl_matrix *k );
```

This function executes the algorithm described in the Section 7.9, given a structure $SubSeqNode$ $s$ and an intrinsic parameter $k$.

```
void subseq_dlt_calib( SubSeqNode *s, gsl_matrix *k );
```

This function is part of the function *subseq_extrinsic_calibrate*. It executes the algorithm of Section 7.3.

```
void subseq_extrinsic_ba( SubSeqNode *s, gsl_matrix *k );
```

This function is part of the function *subseq_extrinsic_calibrate*. It executes a bundle adjustment in the structure $SubSeqNode$ $s$.

```
void camera_set( Camera *c, gsl_matrix *k, gsl_matrix *r,
                 gsl_vector *t );
```

This is an auxiliary function that allocates a structure $Camera$ $c$, given the parameters $k[r|t]$.

```
void subseq_recalibrate( SubSeqNode *s, double inlier_tol );
```

This function is part of the function *subseq_extrinsic_calibrate*. It discards the points that present an error larger than *inlier_tol*.

```
void merge_outliers( int *out_inliers, int *inliers1,
                     int *inliers2, int nfeatures );
```

This is an auxiliary function that receives two inliers arrays, with $nfeatures$ elements, encoded using 1 as a inlier and 0 as an outlier. It returns in $out\_inliers$ as array that has a 1 when both vectors have 1 in the same position and 0 otherwise.

```
int get_ninliers( int *inliers, int n );
```

This function returns the number of inliers encoded in the array $inliers$ of $n$ elements.

```
void copy_projs_inliers( gsl_matrix *dst, gsl_matrix *src,
                         int *inliers );
```

This function copies the inliers rows of the matrix $src$, which represents a list of 2D projections, in the matrix $dst$. The array $inliers$ defines the inliers representing with the value 1 and the outlier with the value 0.

```
void copy_points_inliers( gsl_matrix *dst, gsl_matrix *src,
                          int *inliers );
```

This function copies the inliers rows of the matrix $src$, which represents a list of 3D points, in the matrix $dst$. The array $inliers$ defines the inliers representing the inlier with the value 1 and the outlier with the value 0.

```
void  outliers_eliminate( SubSeqNode *s , int *inliers,
                          int ninliers );
```

This function eliminates the outliers points and projections of the structure $SubSeqNode$ $s$. The array $inliers$ define the inliers in the same way of the previous function, and $ninliers$ is the number of elements of this array.

```
void get_scale_vector( double *scale_vector,
                       CalibSequenceContext *cc );
```

This function returns in the array $scale\_vector$ all the scale factor, between two consecutive structures $SubSeqNode$ of the $CalibSequenceContext$ $c$. It is done following the algorithm described in the Section 7.12.3.

### 7.15.8 Sequence of Frames Code

```
// sequence/sequence.h

#ifndef SEQUENCE_H
#define SEQUENCE_H
```

```
#include <string.h>
#include <gsl/gsl_min.h>
#include "featureslist.h"
#include "ransac.h"
#include "ba.h"

#define RANSAC_REPROJ_TOL 15.
#define RANSAC_NITERS 60
#define FRAGMENT_NFRAMES 60
#define MIN_MOV_DIST 20.
#define MIN_MOV_DIST2 4.
#define INTRISIC_DIST_TOL 100
#define INTERFRAME_REPROJ_TOL 20.
#define MAX_INTERFRAG_NINLIERS 500
#define MAX_NUMBER_OF_SUBSEQ 100

#define   PROJS(S,J)   (S)->fp[ J ].projs
#define   WPROJS(S,J)  (S)->wfp[ J ].projs

typedef struct Camera{
  gsl_matrix *k, *r;
  gsl_vector *t;
}Camera;


typedef struct FrameProjs{
   gsl_matrix *projs;
}FrameProjs;


typedef struct SubSeqNode{
  struct CalibSequenceContext *cc;
  struct SubSeqNode *next;
  int nframes;                       /* Number of frames */
  int nfeatures;                     /* Number of features per frame */
  int first_frame, last_frame;       /* Interval in the FeatureList */
  int *inliers;
  gsl_matrix *points;                /* 3d points */
  FrameProjs *fp;                    /* Projections */
  int *winliers;      /* Work Inliers */
  gsl_matrix *wpoints;               /* Work 3d points */
  FrameProjs *wfp;                   /* Work projections */
  int wnfeatures;                    /* Number of Work Features */
  Camera *cl;
}SubSeqNode;


typedef struct CalibSequenceContext{
  FeaturesList *fl;
  SubSeqNode *sl;
  Camera *cl;
  int nframes;
  int nfeatures;
  int ransac_iterations;
  double ransac_inliers_tol;
  double reproj_inliers_tol;
}CalibSequenceContext;


/* constructors & destructors */
void camera_fields_alloc( Camera *cam );
void camera_fields_dispose( Camera *cam );

CalibSequenceContext *calib_context_alloc( FeaturesList *fl );
```

```
void calib_context_dispose( CalibSequenceContext *cc );

SubSeqNode *subseq_node_alloc( CalibSequenceContext *cc, FeaturesList *fl );
void subseq_node_dispose(  SubSeqNode *sn );

void camera_set( Camera *c, gsl_matrix *k, gsl_matrix *r, gsl_vector *t );

/* calib operation */
void seq_extrinsic_calibrate( CalibSequenceContext *cc, gsl_matrix *k  );

/* sub sequence operations */
void subseq_extrinsic_calibrate( SubSeqNode *s, gsl_matrix *k );
void subseq_dlt_calib( SubSeqNode *s, gsl_matrix *k );
void subseq_extrinsic_ba( SubSeqNode *s, gsl_matrix *k );
void subseq_recalibrate( SubSeqNode *s, double inlier_tol );

/* inliers & outliers operations */
void merge_outliers( int *out_inliers, int *inliers1, int *inliers2,
          int nfeatures );
int get_ninliers( int *inliers, int n );
void copy_projs_inliers( gsl_matrix *dst, gsl_matrix *src, int *inliers );
void copy_points_inliers( gsl_matrix *dst, gsl_matrix *src, int *inliers );
void  outliers_eliminate( SubSeqNode *s , int *inliers, int ninliers );

/* scale adjust function */
void get_scale_vector( double *scale_vector, CalibSequenceContext *cc );

/* misc */
void extract_rt( gsl_matrix *r, gsl_vector *t, gsl_matrix *ref );

#endif

// sequence/subseq_alloc.c

static int get_n_tracked_features( FeaturesList *fl );


SubSeqNode *subseq_node_alloc( CalibSequenceContext *cc, FeaturesList *fl )
{
 int i, j, k, n_tracked_features;
 SubSeqNode *sn;
 Features *aux;

 n_tracked_features = get_n_tracked_features( fl );
 sn = NEWSTRUCT( SubSeqNode );
 sn->cc = cc;
 sn->nframes = fl->nframes;
 sn->nfeatures = n_tracked_features;
 sn->first_frame = fl->head->frame;
 sn->last_frame = fl->tail->frame;
 sn->inliers = NEWTARRAY( cc->nfeatures, int );
 sn->points = gsl_matrix_alloc( n_tracked_features, 3 );
 sn->fp = NEWTARRAY( fl->nframes, FrameProjs );
 sn->winliers = NEWTARRAY( cc->nfeatures, int );
 sn->wpoints = NULL;
 sn->wfp = NEWTARRAY( fl->nframes, FrameProjs );
 sn->wnfeatures = 0;
 sn->cl = NEWTARRAY( fl->nframes, Camera );

 for( aux = fl->head, i = 0; aux != NULL; aux = aux->next, i++ ){
   k = 0;
   sn->wfp[i].projs = NULL;
   sn->fp[i].projs = gsl_matrix_alloc( n_tracked_features, 2 );
   camera_fields_alloc( &sn->cl[i] );
```

```
    for( j = 0; j < cc->nfeatures; j++ ){
      if( fl->tail->status[j] == FEATURE_TRACKED ){
        gsl_matrix_set( sn->fp[i].projs, k, 0, aux->p[j].x );
        gsl_matrix_set( sn->fp[i].projs, k, 1, aux->p[j].y );
        k++;
        sn->inliers[j] = INLIER;
      }
      else
        sn->inliers[j] = OUTLIER;
    }
  }

  sn->next = NULL;
  return sn;
}


void camera_fields_alloc( Camera *cam )
{
  cam->k = gsl_matrix_alloc( 3, 3 );
  cam->r = gsl_matrix_alloc( 3, 3 );
  cam->t = gsl_vector_alloc( 3 );
}


void camera_fields_dispose( Camera *cam )
{
  gsl_matrix_free( cam->k );
  gsl_matrix_free( cam->r );
  gsl_vector_free( cam->t );
}


void subseq_node_dispose(  SubSeqNode *sn )
{
  int i;

  for( i = 0; i < sn->nframes; i++ ){
     camera_fields_dispose( &sn->cl[i] );
     gsl_matrix_free( sn->fp[i].projs );
  }

  gsl_matrix_free( sn->points );
  free( sn->fp );
  free( sn->cl );
  free( sn );
}


int get_n_tracked_features( FeaturesList *fl )
{
  int i, k=0;

  for( i = 0; i < fl->tail->nfeatures; i++ )
    if( fl->tail->status[i] == FEATURE_TRACKED )
      k++;

  return k;
}


// sequence/seq_alloc.c

CalibSequenceContext *calib_context_alloc( FeaturesList *fl )
```

```
{
 CalibSequenceContext *cc;

 cc = NEWSTRUCT( CalibSequenceContext );
 cc->fl = fl;
 cc->sl = NULL;
 cc->cl = NULL;
 cc->nframes = fl->nframes;
 cc->nfeatures = fl->head->nfeatures;
 cc->ransac_iterations = RANSAC_NITERS;
 cc->ransac_inliers_tol =  .005 ;
 cc->reproj_inliers_tol = RANSAC_REPROJ_TOL ;

 return cc;
}

// sequence/sec.c

static void seq_subseq_list_create(  CalibSequenceContext *cc );
static void seq_insert_subseq(  CalibSequenceContext *cc, SubSeqNode *s );
static void seq_adjust_cameras(  CalibSequenceContext *cc  );


void seq_extrinsic_calibrate( CalibSequenceContext *cc, gsl_matrix *k  )
{
 SubSeqNode *aux;

 seq_subseq_list_create( cc  );

 printf( "Video Fragments\n" );
 for(  aux = cc->sl; aux != NULL; aux = aux->next )
   printf( "Frame: %i to %i nfeat = %i\n", aux->first_frame,
           aux->last_frame, aux->nfeatures );
 printf( "Calibration... The last fragment will be discarded\n");
 for(  aux = cc->sl; (aux != NULL) && (aux->next != NULL); aux = aux->next ){
     printf( "Frame: %i to %i\n", aux->first_frame, aux->last_frame );
     subseq_extrinsic_calibrate( aux, k  );
 }
 seq_adjust_cameras( cc );
}


void seq_subseq_list_create(  CalibSequenceContext *cc )
{
 SubSeqNode *aux;
 FeaturesList *fl;

 features_list_begin_frag( cc->fl );
 do{
    seq_insert_subseq( cc,  subseq_node_alloc( cc,
                       fl = features_list_get_next_frag( cc->fl,
                            FRAGMENT_NFRAMES, MIN_MOV_DIST ) ));
    features_list_dispose( fl );
 }while( !features_list_end_frag() );
}


void seq_insert_subseq(  CalibSequenceContext *cc, SubSeqNode *s )
{
 static SubSeqNode *tail = NULL;

 if( tail == NULL ){
   tail = s;
   cc->sl = s;
```

```
  }
  else{
    tail->next = s;
    tail = s;
  }
}


void seq_adjust_cameras(  CalibSequenceContext *cc  )
{
 double scale = 1., scale_vector[MAX_NUMBER_OF_SUBSEQ];
 int i, n, j = 0;
 SubSeqNode *aux;
 gsl_matrix *id, *ref_a1, *ref_a2, *ref_b1, *ref_b2;

 ref_a1 = gsl_matrix_alloc( 3, 4 );
 ref_a2 = gsl_matrix_alloc( 3, 4 );
 ref_b1 = gsl_matrix_alloc( 3, 4 );
 ref_b2 = gsl_matrix_alloc( 3, 4 );
 id = gsl_matrix_alloc( 3,3 );
 gsl_matrix_set_identity(id);

 get_scale_vector( scale_vector, cc );
 for( aux = cc->sl; (aux != NULL) && (aux->next != NULL); aux = aux->next, j++ ){
   n = aux->nframes - 1;
   scale *= scale_vector[j];
   gsl_vector_scale( aux->next->cl[0].t , scale );
   calib_pmatrix_make( ref_a1, id, aux->next->cl[0].r, aux->next->cl[0].t );
   gsl_vector_scale( aux->next->cl[0].t , 1./scale );
   calib_pmatrix_make( ref_a2, id, aux->cl[n].r, aux->cl[n].t );
   for( i = 0; i < aux->next->nframes; i++ ){
      gsl_vector_scale( aux->next->cl[i].t , scale );
      calib_pmatrix_make( ref_b1, id, aux->next->cl[i].r, aux->next->cl[i].t );
      calib_change_ref( ref_a1, ref_a2, ref_b1, ref_b2  );
      extract_rt( aux->next->cl[i].r, aux->next->cl[i].t, ref_b2 );
   }
 }

 gsl_matrix_free( id );
 gsl_matrix_free( ref_a1 );
 gsl_matrix_free( ref_a2 );
 gsl_matrix_free( ref_b1 );
 gsl_matrix_free( ref_b2 );
}


void extract_rt( gsl_matrix *r, gsl_vector *t, gsl_matrix *ref )
{
 int i, j;

 for( i=0; i<3; i++ )
   for( j=0; j<3; j++ )
     gsl_matrix_set( r, i, j, gsl_matrix_get( ref, i, j ) );

  gsl_matrix_get_col( t, ref, 3 );
}


void calib_context_dispose( CalibSequenceContext *cc )
{
 features_list_dispose( cc->fl );
 free( cc );
}
```

```
// sequence/subsec.c

void subseq_extrinsic_calibrate( SubSeqNode *s, gsl_matrix *k )
{
 subseq_dlt_calib( s, k );
 ba_set_lm_max_iterations( 30 );
 subseq_extrinsic_ba( s, k );
 subseq_recalibrate( s, 10. );
 ba_set_lm_max_iterations( 30 );
 subseq_extrinsic_ba( s, k );
 subseq_recalibrate( s, 5. );
 ba_set_lm_max_iterations( 30 );
 subseq_extrinsic_ba( s, k );
 subseq_recalibrate( s, 3. );
 ba_set_lm_max_iterations( 50 );
 subseq_extrinsic_ba( s, k );
}


// sequence/dlt.c

static int  reproj_inliers( int *out_inliers, int *in_inliers, int n_in_inliers,
                            SubSeqNode *s, gsl_matrix *k);
static int is_intrinsic_inlier( gsl_matrix *p, gsl_matrix *k, double tol );
static double matrix33_distancy( gsl_matrix *k, gsl_matrix *kref );
static void reconstruct( SubSeqNode *s, gsl_matrix *k );
static double projs_distancy( SubSeqNode *s, int frame1, int frame2 );


void subseq_dlt_calib( SubSeqNode *s, gsl_matrix *k )
{
 int i;
 gsl_vector *t;
 gsl_matrix  *k_real, *r, *p;

 t = gsl_vector_alloc( 3 );
 r = gsl_matrix_alloc( 3, 3 );
 k_real = gsl_matrix_alloc( 3, 3 );
 p = gsl_matrix_alloc( 3, 4 );

 reconstruct( s, k );
 for(  i = 0; i < s->nframes; i++ ){
   calib_singlecam_dlt(  p,  s->wpoints, WPROJS( s, i ) );
   calib_camfactor( k_real, r, t,  p );
   camera_set( &(s->cl[i]) , k, r, t );
 }

 gsl_vector_free( t );
 gsl_matrix_free( r );
 gsl_matrix_free( k_real );
 gsl_matrix_free( p );
}


void reconstruct( SubSeqNode *s, gsl_matrix *k )
{
  int i, j, frame1, frame2, ninliers, ransac_ninliers, max_ninliers = -1;
  int *inliers, *ransac_inliers, *max_inliers;
  gsl_matrix *max_points;

  max_points = gsl_matrix_alloc(  s->nfeatures ,3 );
  inliers = (int*)malloc( s->nfeatures*sizeof(int) );
  max_inliers = (int*)malloc( s->nfeatures*sizeof(int) );
```

```
  ransac_inliers = (int*)malloc( s->nfeatures*sizeof(int) );

  for( i = s->nframes - 1; i > 0; i-- ){
    for( j = 0; j < i; j++ ) {
      if( projs_distancy( s, i, j ) > MIN_MOV_DIST2 ){

        ransac_ninliers = ransac_reconstruct( ransac_inliers,
                              s->points, PROJS( s,i ), PROJS( s,j ),
                              k, s->cc->ransac_iterations,
                              s->cc->ransac_inliers_tol  );

        if( (ransac_ninliers > 5) && (ransac_ninliers > max_ninliers) ){
           ninliers = reproj_inliers( inliers, ransac_inliers,
           ransac_ninliers, s, k );

           if( ninliers > max_ninliers ){
             max_ninliers = ninliers;
     memcpy( max_inliers, inliers, s->nfeatures * sizeof( int ) );
     gsl_matrix_memcpy( max_points , s->points );
             frame1 = i;
             frame2 = j;
           }
        }
        printf( "(%i, %i)  r:%i nin:%i, max: ( %i, %i )%i\n",
                i, j, ransac_ninliers, ninliers, frame1, frame2, max_ninliers  );
     }
    }
 }

 gsl_matrix_memcpy( s->points, max_points );
 outliers_eliminate( s , max_inliers, max_ninliers );

 gsl_matrix_free( max_points );
 free( inliers );
 free( max_inliers );
 free( ransac_inliers );
}


int reproj_inliers( int *out_inliers, int *in_inliers, int in_ninliers,
                    SubSeqNode *s, gsl_matrix *k )
{
 int i, *proj_inliers, out_ninliers;
 gsl_matrix *projs, *reprojs, *inline_points, *p;

 p = gsl_matrix_alloc( 3, 4 );
 projs = gsl_matrix_alloc( in_ninliers, 2 );
 reprojs = gsl_matrix_alloc( in_ninliers, 2 );
 inline_points =  gsl_matrix_alloc( in_ninliers, 3 );
 proj_inliers = (int*)malloc( in_ninliers*sizeof(int) );

 copy_points_inliers( inline_points, s->points, in_inliers );
 memcpy( out_inliers, in_inliers, s->nfeatures * sizeof( int ) );
 for( i = 0; i < s->nframes; i++ ) {
   copy_projs_inliers( projs, PROJS( s, i ), in_inliers );
   ransac_singlecam( proj_inliers, p, inline_points, projs,
                     s->cc->ransac_iterations, s->cc->reproj_inliers_tol );
   merge_outliers( out_inliers, in_inliers, proj_inliers, s->nfeatures );
 }

 out_ninliers = get_ninliers( out_inliers, s->nfeatures );
 if( out_ninliers > 5 ){
   gsl_matrix_free( projs );
   gsl_matrix_free( inline_points );
```

```
      projs = gsl_matrix_alloc( out_ninliers, 2 );
      inline_points =  gsl_matrix_alloc( out_ninliers, 3 );
      copy_points_inliers( inline_points, s->points, out_inliers );
      for( i = 0; i < s->nframes; i++ ){
        copy_projs_inliers( projs, PROJS( s, i ), out_inliers );
        calib_singlecam_dlt( p, inline_points, projs );
        if( !is_intrinsic_inlier(p, k, INTRISIC_DIST_TOL) ){
          out_ninliers = -1;
          break;
        }
      }
    }

  gsl_matrix_free( p );
  gsl_matrix_free( projs );
  gsl_matrix_free( reprojs );
  gsl_matrix_free( inline_points );
  free( proj_inliers );
  return out_ninliers;
}


int is_intrinsic_inlier( gsl_matrix *p, gsl_matrix *kref, double tol )
{
 double d;
 gsl_matrix *k, *r;
 gsl_vector *t;

 k = gsl_matrix_alloc( 3, 3 );
 r = gsl_matrix_alloc( 3, 3 );
 t = gsl_vector_alloc( 3 );

 calib_camfactor( k, r, t, p );
 d = matrix33_distancy( k, kref );

 gsl_matrix_free( k );
 gsl_matrix_free( r );
 gsl_vector_free( t );

 if( d < tol )
   return 1;
 else
   return 0;
}


#ifndef MAX
  #define MAX(U,V) (U>V?U:V)
#endif

double matrix33_distancy( gsl_matrix *k, gsl_matrix *kref )
{
 double e = 0;
 int i, j;

 for( i = 0; i < 3; i++ )
   for( j = 0; j < 3; j++ )
     e = MAX( e, fabs( gsl_matrix_get(k, i, j )- gsl_matrix_get( kref, i, j )) );

 return e;
}


double projs_distancy( SubSeqNode *s, int frame1, int frame2 )
```

```
{
 gsl_vector *v1, *v2;
 int i, ninliers = 0;
 double d = 0;

 v1 = gsl_vector_alloc(2);
 v2 = gsl_vector_alloc(2);
 for( i = 0; i < s->nfeatures; i++ ){
   gsl_matrix_get_row( v1, PROJS(s,frame1), i );
   gsl_matrix_get_row( v2, PROJS(s,frame2), i );
   gsl_vector_sub( v1, v2 );
   d += gsl_blas_dnrm2( v1 );
 }

 gsl_vector_free( v1 );
 gsl_vector_free( v2 );
 return  d/s->nfeatures;
}


// sequence/ba.c

static void  set_frame_projs( gsl_matrix *projs_buffer,
                              gsl_matrix *projs, int cam_index );
static void set_all_points( gsl_vector *params, gsl_matrix *points );
static void get_all_points( gsl_matrix *points, gsl_vector *params );


void subseq_extrinsic_ba( SubSeqNode *s, gsl_matrix *k )
{
 int i;
 gsl_vector *xout, *params, *t, *raxis;
 gsl_matrix  *projs, *r;

 xout = ba_ext_param_alloc( s->nframes, s->wnfeatures  );
 params = ba_ext_param_alloc( s->nframes, s->wnfeatures  );
 projs = gsl_matrix_alloc( s->wnfeatures,  2*s->nframes );
 raxis = gsl_vector_alloc( 3 );
 r = gsl_matrix_alloc( 3, 3 );
 t = gsl_vector_alloc( 3 );

 printf( "Bundle Adjustment...\n" );
 set_all_points( params, s->wpoints );
 for( i = 0; i < s->nframes; i++ ){
    ba_r_to_axis_angle( raxis,  s->cl[i].r  );
    ba_ext_set_camera( params, raxis, s->cl[i].t, s->wnfeatures, i );
    set_frame_projs( projs, WPROJS( s, i ), i );
 }

 ba_ext_exec( xout, params, projs , k );

 for( i = 0; i < s->nframes; i++ ){
      ba_ext_get_rt( raxis, t, xout, s->wnfeatures,  i );
      ba_axis_angle_to_r( r, raxis );
      camera_set( &(s->cl[i]), k,  r, t );
 }
 get_all_points( s->wpoints, params );

 gsl_vector_free( xout );
 gsl_vector_free( params );
 gsl_matrix_free( projs );
 gsl_vector_free( raxis );
 gsl_matrix_free( r );
 gsl_vector_free( t );
```

```
}


void  set_frame_projs( gsl_matrix *projs_buffer,  gsl_matrix *projs, int cam_index )
{
 int i;
 gsl_vector *proj;

 proj = gsl_vector_alloc( 2 );
 for( i = 0; i < projs->size1; i++ ){
    gsl_matrix_get_row(  proj, projs, i );
    ba_set_proj( projs_buffer, proj, cam_index, i );
 }

 gsl_vector_free( proj );
}


void set_all_points( gsl_vector *params, gsl_matrix *points )
{
 int i;
 gsl_vector *v;

 v = gsl_vector_alloc( 3 );
 for( i = 0; i < points->size1; i++ ){
   gsl_matrix_get_row( v, points, i );
   ba_set_point( params, v, i );
 }

 gsl_vector_free( v );
}


void get_all_points( gsl_matrix *points, gsl_vector *params )
{
 int i;
 gsl_vector *v;

 v = gsl_vector_alloc( 3 );
 for( i = 0; i < points->size1; i++ ){
   ba_get_point( v, params, i );
   gsl_matrix_set_row( points, i,  v );
 }

 gsl_vector_free( v );
}


void camera_set( Camera *c, gsl_matrix *k, gsl_matrix *r, gsl_vector *t )
{
 gsl_matrix_memcpy( c->k, k );
 gsl_matrix_memcpy( c->r, r );
 gsl_vector_memcpy( c->t, t );
}


// sequence/recalibrate.c

static int is_inlier( SubSeqNode *s, gsl_vector *x, int index, double tol );

void subseq_recalibrate( SubSeqNode *s, double inlier_tol )
{
 int i,j,k, ninliers = 0;
 int *inliers;
```

```
 gsl_vector *x, *x1, *x2;
 gsl_matrix *p, *p1, *p2;

 inliers = NEWTARRAY( s->nfeatures, int );
 x = gsl_vector_alloc(3);
 x1 = gsl_vector_alloc(2);
 x2 = gsl_vector_alloc(2);
 p = gsl_matrix_alloc(3,4);
 p1 = gsl_matrix_alloc(3,4);
 p2 = gsl_matrix_alloc(3,4);

 printf( "Recalibrating...\nInlier tol = %f pixel\n", inlier_tol );
 for( k = 0; k < s->nfeatures; k++ ){
   inliers[k] = OUTLIER;
   for( i = 0; i < s->nframes; i++ )
     for( j = 0; j < i; j++ ){
       calib_pmatrix_make( p1 , s->cl[i].k, s->cl[i].r, s->cl[i].t );
       calib_pmatrix_make( p2 , s->cl[j].k, s->cl[j].r, s->cl[j].t );
       gsl_matrix_get_row( x1, PROJS( s, i ), k );
       gsl_matrix_get_row( x2, PROJS( s, j ), k );
       calib_get_3dpoint( p1, p2, x1, x2, x );
       if( is_inlier( s, x, k, inlier_tol ) && (inliers[k] != INLIER) ){
         gsl_matrix_set_row( s->points, k, x );
         inliers[k] = INLIER;
         ninliers++;
         i = j = s->nframes;
       }
     }
 }
 outliers_eliminate( s, inliers, ninliers );
 printf( "Number of inliers: %i\n", ninliers );

 free( inliers );
 gsl_vector_free( x );
 gsl_vector_free( x1 );
 gsl_vector_free( x2 );
 gsl_matrix_free( p );
 gsl_matrix_free( p1 );
 gsl_matrix_free( p2 );
}


int is_inlier( SubSeqNode *s, gsl_vector *x, int index, double tol )
{
 int i, n = 0;
 gsl_vector *x_proj, *x_real;
 gsl_matrix *p;

 x_proj = gsl_vector_alloc(2);
 x_real = gsl_vector_alloc(2);
 p = gsl_matrix_alloc(3, 4);

 for( i = 0; i < s->nframes; i++ ){
    calib_pmatrix_make( p , s->cl[i].k, s->cl[i].r, s->cl[i].t );
    calib_apply_P( p, x, x_proj );
    gsl_matrix_get_row( x_real, PROJS( s, i ), index );
    gsl_vector_sub( x_proj, x_real );
    if( gsl_blas_dnrm2( x_proj ) < tol )
      n++;
 }

 gsl_vector_free( x_proj );
 gsl_vector_free( x_real );
 gsl_matrix_free( p );
```

```
 if( n == s->nframes )
   return TRUE;
 else
   return FALSE;
}


// sequence/inliers.c

void merge_outliers( int *out_inliers, int *inliers1, int *inliers2, int nfeatures )
{
 int j, k = 0;

 for( k = 0, j = 0; j < nfeatures ; j++ ){
   if( inliers1[j] == INLIER ){
     if( inliers2[k] == OUTLIER )
         out_inliers[j] = OUTLIER;
     k++;
   }
 }
}


int get_ninliers( int *inliers, int n )
{
 int i, ninliers = 0;

 for( i = 0; i < n; i++ )
   if( inliers[i] == INLIER )
     ninliers++;

 return  ninliers;
}


void copy_projs_inliers( gsl_matrix *dst, gsl_matrix *src, int *inliers )
{
 int i,  k=0;

 for( i = 0; i < src->size1; i++ )
   if( inliers[i] == INLIER ){
     gsl_matrix_set( dst, k, 0, gsl_matrix_get( src, i , 0 ) );
     gsl_matrix_set( dst, k, 1, gsl_matrix_get( src, i , 1 ) );
     k++;
   }
}


void copy_points_inliers( gsl_matrix *dst, gsl_matrix *src, int *inliers )
{
 int i,  k=0;

 for( i = 0; i < src->size1; i++ )
   if( inliers[i] == INLIER ){
     gsl_matrix_set( dst, k, 0, gsl_matrix_get( src, i , 0 ) );
     gsl_matrix_set( dst, k, 1, gsl_matrix_get( src, i , 1 ) );
     gsl_matrix_set( dst, k, 2, gsl_matrix_get( src, i , 2 ) );
     k++;
   }
}


void  outliers_eliminate( SubSeqNode *s , int *inliers, int ninliers )
```

```
{
 int i, k = 0;
 gsl_matrix *projs, *points;

 for( i = 0; i < s->nframes; i++ ) {
   projs = gsl_matrix_alloc( ninliers, 2 );
   copy_projs_inliers( projs, PROJS( s, i ), inliers );
   if( WPROJS(s,i) != NULL )
     gsl_matrix_free( WPROJS(s,i) );
   WPROJS( s, i ) = projs;
 }

for( i = 0; i < s->nfeatures; i++ ){
   while( s->inliers[k] == OUTLIER )
     s->winliers[k++] = OUTLIER;
   if( inliers[i] == INLIER )
     s->winliers[k] = INLIER;
   else
     s->winliers[k] = OUTLIER;
   k++;
 }

 s->wnfeatures = ninliers;
 points =  gsl_matrix_alloc( ninliers, 3 );
 copy_points_inliers( points, s->points, inliers );
 if( s->wpoints != NULL )
    gsl_matrix_free( s->wpoints );
 s->wpoints = points;
}


// sequence/scale.c

#define SCALE_MAX_ITERATIONS 50
#define SCALE_OPTIMIZE_RADIUS 1.01
#define MIN_ACCEPTABLE_NINLIERS 3


typedef struct ScaleOptimizeContext{
  int *reproj_inliers;
  int n;  /* p1 and p2 number of elements */
  int *p1;
  int *p2;
  SubSeqNode *subseq;
}ScaleOptimizeContext;


static double get_scale( SubSeqNode *s );
static int get_good_vector( int *good, SubSeqNode *s );
static void get_inlier_index( int *v , int *inlier, int *good, int n);
static int scale_reproj_inliers( int *reproj_inliers, gsl_vector *error,
                              double scale, SubSeqNode *s, int *p1, int *p2 );
static double scale_optimize( int *reproj_inliers, double scale, SubSeqNode *s,
                            int *p1, int *p2, int n_interframe_inliers );
static double scale_cost_func( double x, void *params );
static double mean_global_error( gsl_vector *error, int *reproj_inliers );


void get_scale_vector( double *scale_vector, CalibSequenceContext *cc )
{
 int i = 0;
 SubSeqNode *aux;

 for( aux = cc->sl; (aux != NULL) && (aux->next != NULL) &&
```

```
      (aux->next->next != NULL); aux = aux->next )
         scale_vector[i++] = get_scale(aux);
}


double get_scale( SubSeqNode *s )
{
 double scale, best_scale = 1;
 double global_error, min_global_error = MAX_DOUBLE;
 int best_ninliers, ninliers, i, k;
 int good[MAX_INTERFRAG_NINLIERS];
 int p1[MAX_INTERFRAG_NINLIERS];
 int p2[MAX_INTERFRAG_NINLIERS];
 int *reproj_inliers, *best_inliers;
 gsl_vector *error;

 k = get_good_vector( good, s );
 get_inlier_index( p1, s->winliers, good, s->cc->nfeatures );
 get_inlier_index( p2, s->next->winliers, good, s->cc->nfeatures);
 reproj_inliers = NEWTARRAY( k, int );
 best_inliers = NEWTARRAY( k, int );
 error = gsl_vector_alloc(k);

 printf( "Shared Inliers %i\n", k );
 printf( "Frame: %i to %i\n", s->first_frame, s->last_frame );
 for( scale = EPS; scale < 100.; scale *= SCALE_OPTIMIZE_RADIUS ){
    ninliers = scale_reproj_inliers( reproj_inliers, error, scale, s, p1, p2 );
    if( (global_error = mean_global_error( error, reproj_inliers ))
    <= min_global_error ){
          best_ninliers = ninliers;
          best_scale = scale;
          min_global_error = global_error;
          memcpy( best_inliers, reproj_inliers, k*sizeof(int) );
    }
 }
 printf( "best scale: %f , ninliers = %i, error = %f\n", best_scale, best_ninliers,
          min_global_error/s->nframes );
 best_scale = scale_optimize( best_inliers, best_scale, s, p1, p2, k );

 free( reproj_inliers );
 free( best_inliers );
 gsl_vector_free( error );
 return best_scale;
}


static int get_good_vector( int *good, SubSeqNode *s )
{
 int i, k = 0;

 for( i = 0; i < s->cc->nfeatures; i++ )
    if( s->winliers[i] == INLIER && s->next->winliers[i] == INLIER ){
      good[k++] = i;
    }
 return k;
}


static void get_inlier_index( int *v , int *inlier, int *good, int n )
{
 int i, j=0, k=0, w = 0;

 for( i = 0; i < n; i++ ){
   if( inlier[i] == INLIER ){
```

```
      if( i == good[j] )
        v[j++] = k;
      k++;
    }
  }
}


static int scale_reproj_inliers( int *reproj_inliers, gsl_vector *error,
                                 double scale, SubSeqNode *s, int *p1, int *p2 )
{
 int i,j, n, w, ninliers = 0;
 gsl_vector *v, *v_proj, *real_proj, *t;
 gsl_matrix *id, *r, *ref_a1, *ref_a2, *ref_b1, *ref_b2, *p;

 v = gsl_vector_alloc( 3 );
 v_proj = gsl_vector_alloc( 2 );
 real_proj = gsl_vector_alloc( 2 );
 t = gsl_vector_alloc( 3 );
 id = gsl_matrix_alloc( 3,3 );
 r = gsl_matrix_alloc( 3, 3 );
 ref_a1 = gsl_matrix_alloc( 3, 4 );
 ref_a2 = gsl_matrix_alloc( 3, 4 );
 ref_b1 = gsl_matrix_alloc( 3, 4 );
 ref_b2 = gsl_matrix_alloc( 3, 4 );
 p = gsl_matrix_alloc( 3, 4 );

 gsl_matrix_set_identity(id);
 n = s->nframes - 1;
 gsl_vector_memcpy( t, s->next->cl[0].t );
 gsl_vector_scale( t, scale );
 calib_pmatrix_make( ref_a1, id, s->next->cl[0].r, t );
 calib_pmatrix_make( ref_a2, id, s->cl[n].r, s->cl[n].t );

 for( i = 0; i < error->size; i++ ){
  w = 0;
  gsl_vector_set( error, i, 0. );
  gsl_matrix_get_row( v, s->wpoints, p1[i] );
  for( j = 0; j < s->next->nframes; j++ ){
     gsl_vector_memcpy( t, s->next->cl[j].t );
     gsl_vector_scale( t, scale );
     calib_pmatrix_make( ref_b1, id, s->next->cl[j].r, t );
     calib_change_ref( ref_a1, ref_a2, ref_b1, ref_b2  );
     extract_rt( r, t, ref_b2 );
     calib_pmatrix_make( p, s->next->cl[j].k, r, t );
     calib_apply_P( p, v, v_proj );
     gsl_matrix_get_row( real_proj, WPROJS( s->next, j ), p2[i] );
     gsl_vector_sub( v_proj, real_proj );
     gsl_vector_set( error, i, gsl_blas_dnrm2( v_proj ) +
                     gsl_vector_get(error,i) );
     if( gsl_blas_dnrm2( v_proj )  < INTERFRAME_REPROJ_TOL )
        w++;
   }
   if( w == s->next->nframes ){
     ninliers++;
     reproj_inliers[i] = INLIER;
   }
   else
     reproj_inliers[i] = OUTLIER;
 }

 gsl_vector_free( v );
 gsl_vector_free( v_proj );
 gsl_vector_free( real_proj );
```

```
  gsl_vector_free( t );
  gsl_matrix_free( id );
  gsl_matrix_free( r );
  gsl_matrix_free( ref_a1 );
  gsl_matrix_free( ref_a2 );
  gsl_matrix_free( ref_b1 );
  gsl_matrix_free( ref_b2 );
  gsl_matrix_free( p );
  return ninliers;
}


static double scale_optimize( int *reproj_inliers, double scale,
                              SubSeqNode *s, int *p1, int *p2, int n )
{
  int status, iter = 0;
  double a = EPS;
  double b = 100.;
  const gsl_min_fminimizer_type *T;
  gsl_min_fminimizer *m;
  gsl_function F;
  ScaleOptimizeContext sc;

  sc.reproj_inliers = reproj_inliers;
  sc.n = n;
  sc.p1 = p1;
  sc.p2 = p2;
  sc.subseq = s;

  F.function = &scale_cost_func;
  F.params = &sc;

  T = gsl_min_fminimizer_brent;
  m = gsl_min_fminimizer_alloc(T);
  gsl_min_fminimizer_set (m, &F, scale, a, b );

  do{
   iter++;
   status = gsl_min_fminimizer_iterate(m);
   scale = gsl_min_fminimizer_x_minimum(m);
   a = gsl_min_fminimizer_x_lower(m);
   b = gsl_min_fminimizer_x_upper(m);
   status = gsl_min_test_interval (a, b, 1e-10, 0.0);
  }
  while (status == GSL_CONTINUE && iter < SCALE_MAX_ITERATIONS );
  gsl_min_fminimizer_free(m);

  return scale;
}


double scale_cost_func( double x, void *params )
{
  int i,j;
  double global_error = 0;
  int *inliers;
  gsl_vector *error;
  ScaleOptimizeContext *sc;

  sc = (ScaleOptimizeContext*)params;
  inliers = NEWTARRAY( sc->n , int );
  error = gsl_vector_alloc( sc->n );

  scale_reproj_inliers( inliers, error, x, sc->subseq, sc->p1, sc->p2 );
```

```
 global_error = mean_global_error( error, inliers );

 free( inliers );
 gsl_vector_free( error );
 return global_error;
}


double mean_global_error( gsl_vector *error, int *reproj_inliers )
{
 int i, n=0;
 double global_error = 0;

 for( i=0; i < error->size; i++ )
   if( reproj_inliers[i] == INLIER ){
     global_error += gsl_vector_get( error, i );
     n++;
   }

 if( n > MIN_ACCEPTABLE_NINLIERS )
   return global_error/n;
 else
   return MAX_DOUBLE;
}

// calib/changeref.c

#include "calib.h"


void calib_change_ref( gsl_matrix *ref_a1, gsl_matrix *ref_a2,
                       gsl_matrix *ref_b1, gsl_matrix *ref_b2  )
{
 int i,j;
 gsl_matrix *ra1, *ra2, *rb1, *rb2,
            *ta1, *ta2, *tb1, *tb2, *raux, *r, *t;

 r = gsl_matrix_alloc( 3, 3 );
 raux = gsl_matrix_alloc( 3, 3 );
 ra1 = gsl_matrix_alloc( 3, 3 );
 ra2 = gsl_matrix_alloc( 3, 3 );
 rb1 = gsl_matrix_alloc( 3, 3 );
 rb2 = gsl_matrix_alloc( 3, 3 );
 ta1 = gsl_matrix_alloc( 3, 1 );
 ta2 = gsl_matrix_alloc( 3, 1 );
 tb1 = gsl_matrix_alloc( 3, 1 );
 t = gsl_matrix_alloc( 3, 1 );

 for( i=0; i<3; i++ )
   for( j=0; j<3; j++ ){
     gsl_matrix_set( ra1, i, j, gsl_matrix_get( ref_a1, i, j ) );
     gsl_matrix_set( ra2, i, j, gsl_matrix_get( ref_a2, i, j ) );
     gsl_matrix_set( rb1, i, j, gsl_matrix_get( ref_b1, i, j ) );
   }

 for( i=0; i<3; i++ ){
   gsl_matrix_set( ta1, i, 0, gsl_matrix_get( ref_a1, i, 3 ) );
   gsl_matrix_set( ta2, i, 0, gsl_matrix_get( ref_a2, i, 3 ) );
   gsl_matrix_set( tb1, i, 0, gsl_matrix_get( ref_b1, i, 3 ) );
 }

 gsl_matrix_transpose( ra1 );
 gsl_linalg_matmult( rb1, ra1, raux );
 gsl_linalg_matmult( raux, ra2, r );
```

```
    gsl_matrix_sub( ta2, ta1 );
    gsl_linalg_matmult( raux, ta2, t );
    gsl_matrix_add( t, tb1 );

    for( i=0; i<3; i++ )
      for( j=0; j<3; j++ )
        gsl_matrix_set( ref_b2, i, j, gsl_matrix_get( r, i, j ) );

    for( i=0; i<3; i++ )
      gsl_matrix_set( ref_b2, i, 3, gsl_matrix_get( t, i, 0 ) );

    gsl_matrix_free(r);
    gsl_matrix_free(raux);
    gsl_matrix_free(ra1);
    gsl_matrix_free(ra2);
    gsl_matrix_free(rb1);
    gsl_matrix_free(rb2);
    gsl_matrix_free(ta1);
    gsl_matrix_free(ta2);
    gsl_matrix_free(tb1);
    gsl_matrix_free(t);
}
```

### 7.15.9 Relaxation API

```
Cameras *cameras_alloc( gsl_matrix *k,
                        gsl_matrix *r[MAX_CAMERAS],
                        gsl_vector *t[MAX_CAMERAS],
                        int ncams );
```

This function allocates a data structure *Cameras* with the initial values of the extrinsic parameters $[r|t]$ of *ncams* cameras, such that $ncams < MAX\_CAMERAS$. The $k$ matrix must receive the intrinsic parameters of the camera.

```
Cameras *cameras_read( gsl_matrix *k, FILE *f );
```

This function initialize a data structure *Cameras* that represent all cameras whose intrinsic parameters are defined by the matrix $k$, and the extrinsic parameters are listed in the file $f$ compatible with the output of the program *MatchMove* described on Section 7.16.

```
void cameras_free( Cameras *c );
```

This function destroys the data structure *Cameras c*.

```
void cameras_apply( Cameras *c, gsl_vector *p,
                    gsl_vector *proj, int i );
```

This function applies the $i^{th}$ camera of $c$ to the vector $p$ and returns the projection in $proj$.

```
void cameras_getP( gsl_matrix *p, Cameras *c, int i );
```

This function copies in the $3 \times 4$ matrix $p$, the $i^{th}$ camera of $c$.

```
HomologPoint  *hp_alloc( int first_frame, int last_frame );
```

This function allocates a structure for representing all the homologous points from $first\_frame$ to $last\_frame$.

```
void hp_free( HomologPoint *hp );
```

This function deallocates the structure $hp$ used for representing a set of homologous points.

```
Boolean hp_has_frame( HomologPoint *hp, int frame );
```

This function returns $TRUE$ if the frame indexed by $frame$ belongs to the structure $hp$, and $FALSE$ otherwise.

```
Boolean hp_is_inlier( HomologPoint *hp, int frame );
```

This function returns $TRUE$ if the frame indexd by $frame$ is considered inlier in the structure $hp$.

```
void hp_set_inlier( HomologPoint *hp, int frame, Boolean b );
```

This function uses the parameter $b$ in order to define if the homologous points from $hp$ is inlier or outlier in the frame indexed by $frame$.

```
void hp_set_proj( HomologPoint *hp, int frame,
                  double x, double y );
```

This function sets to $(x, y)$ the coordinates of the homologous point described by the data structure $hp$, in the frame indexed by $frame$.

```
void hp_get_proj( gsl_vector *v, HomologPoint *hp,
                  int frame );
```

This function puts in $v$ the coordinates of the homologous point described by the data structure *hp* in the frame indexed by $frame$.

```
void hp_reconstruct_points( HomologPoint *hp, Cameras *c,
                            double tol_error );
```

This function returns, in the field $p$ of *hp*, the best 3D reconstruction of the points of *hp* using a pair of cameras in the reconstruction process, given a set of cameras $c$, and a tolerance error $tol\_error$. The pair of cameras considered in the reconstruction is the one that maximizes the number of inliers.

```
int hp_ninliers( HomologPoint *hp, gsl_vector *point,
      Boolean *inliers, Cameras *c, double tol_error );
```

This function returns the number of frames in which the *point* is considered an inlier point assuming the *tol_error* tolerance. The parameter $c$ receives the set of cameras and *hp* receives the correspondent list of homologous points. The *hp* structure has its inliers field adjusted after the execution of this function.

```
void adjust_inliers( HomologPoint *hp, Boolean *inliers );
```

This function copies the array $inliers$ in the array $hp \rightarrow is\_inlier$.

```
double hp_reproj_error( HomologPoint *hp, int i,
      Cameras *c );
```

This function returns the reprojection error of the point $hp \rightarrow p$ make by the $i^{th}$ camera of $c$, assuming the correspondent list of homologous points passed in *hp*.

```
void relax_cameras( gsl_vector *xout,
               RelaxationContext *rc, int cam_id );
```

This function executes the relaxation of the camera indexed by *cam_id* using the structure $RelaxationContext\ rc$.

```
void adjust_relax_context( RelaxationContext *rc,
                     gsl_vector *xout, int cam_id );
```

This function actualize camera indexed by *cam_id* in the sctructure $RelaxationContext$ $rc$ using the output $xout$ of the function $relax\_cameras$.

```
void relax_points( gsl_vector *xout, gsl_vector *x,
                   HomologPoint *hp, Cameras *c );
```

This function executes the relaxation of the 3D point $hp \rightarrow p$ encoded in the $x$ parameter, using the list of cameras $c$. $xout$ receives the output 3D coordinates.

```
void relax_optimize( gsl_multifit_fdfsolver *s );
```

It is an auxiliary function used for solving the Levenberg-Marquardt in the $GnuScientificLibrary$.

```
RelaxContext *rc_alloc( char *fname, Cameras *c,
      double tol_error );
```

This function allocates a $RelaxContext$ structure. $fname$ is the name of the file generated by the program of Section 6.7. $c$ are the initial cameras estimated by the Match-Move Program, and *tol_error* is threshold used for classifying inliers.

### 7.15.10 Relaxation Code

```
// relax/cameras.h

#ifndef CAMERAS_H
#define CAMERAS_H

#include "defs.h"
#include "calib.h"

#define MAX_CAMERAS 3000

typedef struct Cameras{
  gsl_matrix *k;
  gsl_matrix *r[MAX_CAMERAS];
  gsl_vector *t[MAX_CAMERAS];
  int ncams;
} Cameras;


Cameras *cameras_alloc( gsl_matrix *k, gsl_matrix *r[MAX_CAMERAS],
                        gsl_vector *t[MAX_CAMERAS], int ncams );
Cameras *cameras_read( gsl_matrix *k, FILE *f );
void cameras_free( Cameras *c );
void cameras_apply( Cameras *c, gsl_vector *p, gsl_vector *proj, int i );
void cameras_getP( gsl_matrix *p, Cameras *c, int i );

#endif

// relax/cameras.c

Cameras *cameras_alloc( gsl_matrix *k, gsl_matrix *r[MAX_CAMERAS],
                        gsl_vector *t[MAX_CAMERAS], int ncams )
```

```
{
 int i;
 Cameras *c = NEWSTRUCT(Cameras);

 c->k = k;
 for( i = 0; i < ncams; i++ ){
    c->r[i] = r[i];
    c->t[i] = t[i];
 }
 c->ncams = ncams;

 return c;
}


Cameras *cameras_read( gsl_matrix *k, FILE *f )
{
 int i = 0;
 gsl_matrix *r[MAX_CAMERAS];
 gsl_vector *t[MAX_CAMERAS];

 while( !feof(f) ){
   fscanf( f, "Frame %i\n", &i );
   r[i] = gsl_matrix_alloc(3,3);
   gsl_matrix_fscanf( f, r[i] );
   fscanf( f, "\n");
   t[i] = gsl_vector_alloc(3);
   gsl_vector_fscanf( f, t[i] );
   fscanf( f, "\n");
   i++;
 }

 return cameras_alloc( k, r, t, i );
}


void cameras_free( Cameras *c )
{
 int i;

 gsl_matrix_free(c->k);
 for( i = 0; i < c->ncams; i++ ){
   gsl_matrix_free( c->r[i] );
   gsl_vector_free( c->t[i] );
 }
}


void cameras_apply( Cameras *c, gsl_vector *p, gsl_vector *proj, int i )
{
 gsl_matrix *m;

 m = gsl_matrix_alloc( 3, 4 );
 calib_pmatrix_make( m, c->k, c->r[i], c->t[i] );
 calib_apply_P( m, p, proj );

 gsl_matrix_free(m);
}


void cameras_getP( gsl_matrix *p, Cameras *c, int i )
{
 calib_pmatrix_make( p, c->k, c->r[i], c->t[i] );
}
```

```
// relax/homol_point.h

#ifndef HOMOL_POINT_H
#define HOMOL_POINT_H

#include "cameras.h"
#include "calib.h"

#define MAX_FRAMES_PER_FEATURE 1000

typedef struct HomologPoint{
 int first_frame;
 int last_frame;
 gsl_vector *p;
 gsl_matrix *projs;
 Boolean *is_inlier;
 int ninliers;
 int index;
} HomologPoint;


HomologPoint  *hp_alloc( int first_frame, int last_frame );
void hp_free( HomologPoint *hp );
Boolean hp_has_frame( HomologPoint *hp, int frame );
Boolean hp_is_inlier( HomologPoint *hp, int frame );
void hp_set_inlier(  HomologPoint *hp, int frame, Boolean b );
void hp_set_proj( HomologPoint *hp, int frame, double x, double y );
void hp_get_proj( gsl_vector *v, HomologPoint *hp, int frame );
void hp_reconstruct_points( HomologPoint *hp, Cameras *c, double to_error );
int hp_ninliers( HomologPoint *hp, gsl_vector *point, Boolean *inliers,
                 Cameras *c, double tol_error );
void adjust_inliers( HomologPoint *hp, Boolean *inliers );
double hp_reproj_error( HomologPoint *hp, int i, Cameras *c );

#endif

// relax/homol_point.c

HomologPoint  *hp_alloc( int first_frame, int last_frame )
{
 HomologPoint *hp = NEWSTRUCT(HomologPoint);

 hp->first_frame = first_frame;
 hp->last_frame = last_frame;
 hp->p = gsl_vector_alloc( 3 );
 hp->projs = gsl_matrix_alloc( last_frame - first_frame + 1, 2 );
 hp->is_inlier = NEWTARRAY( last_frame - first_frame + 1, Boolean );
 hp->ninliers = 0;

 return hp;
}


void hp_free( HomologPoint *hp )
{
 gsl_vector_free( hp->p );
 gsl_matrix_free( hp->projs );
 free( hp );
}


Boolean hp_has_frame( HomologPoint *hp, int frame )
{
```

```
  if( (frame >= hp->first_frame ) &&
      (frame <= hp->last_frame ) ){
    return TRUE;
  }
  return FALSE;
}

Boolean hp_is_inlier( HomologPoint *hp, int frame )
{
  if( hp_has_frame( hp, frame ) &&
      hp->is_inlier[frame - hp->first_frame] ){
    return TRUE;
  }
  return FALSE;
}


void hp_set_inlier(  HomologPoint *hp, int frame, Boolean b )
{
  hp->is_inlier[frame - hp->first_frame] = b;
}


void hp_set_proj( HomologPoint *hp, int frame, double x, double y )
{
  gsl_matrix_set( hp->projs, frame - hp->first_frame, 0, x );
  gsl_matrix_set( hp->projs, frame - hp->first_frame, 1, y );
}



void hp_get_proj( gsl_vector *v, HomologPoint *hp, int frame )
{
  gsl_vector_set( v, 0, gsl_matrix_get( hp->projs, frame - hp->first_frame, 0 ) );
  gsl_vector_set( v, 1, gsl_matrix_get( hp->projs, frame - hp->first_frame, 1 ) );
}


void hp_reconstruct_points( HomologPoint *hp, Cameras *c, double tol_error )
{
  double error;
  int i, j, ninliers, max_ninliers = 0;
  gsl_matrix *pi, *pj;
  gsl_vector *pt1, *pt2, *point;
  Boolean inliers[MAX_FRAMES_PER_FEATURE];

  pi = gsl_matrix_alloc( 3, 4 );
  pj = gsl_matrix_alloc( 3, 4 );
  pt1 = gsl_vector_alloc( 2 );
  pt2 = gsl_vector_alloc( 2 );
  point = gsl_vector_alloc( 3 );

  for( i = hp->first_frame; i <= hp->last_frame; i++ ){
     for( j = i+1; j <= hp->last_frame; j++ ){
       cameras_getP(pi, c, i);
       cameras_getP(pj, c, j);

       gsl_vector_set( pt1, 0, gsl_matrix_get( hp->projs, i - hp->first_frame, 0 ) );
       gsl_vector_set( pt1, 1, gsl_matrix_get( hp->projs, i - hp->first_frame, 1 ) );
       gsl_vector_set( pt2, 0, gsl_matrix_get( hp->projs, j - hp->first_frame, 0 ) );
       gsl_vector_set( pt2, 1, gsl_matrix_get( hp->projs, j - hp->first_frame, 1 ) );

       calib_get_3dpoint( pi, pj, pt1, pt2, point );
       ninliers = hp_ninliers( hp, point, inliers, c, tol_error );
```

```
            if( ninliers > max_ninliers ){
               gsl_vector_memcpy( hp->p, point );
               max_ninliers = hp->ninliers = ninliers;
               adjust_inliers( hp, inliers );
            }
         }
     }

     gsl_matrix_free(pi);
     gsl_matrix_free(pj);
     gsl_vector_free(pt1);
     gsl_vector_free(pt2);
     gsl_vector_free(point);
}


int hp_ninliers( HomologPoint *hp, gsl_vector *point, Boolean *inliers,
                 Cameras *c, double tol_error )
{
 double error;
 int i, ninliers = 0;
 gsl_vector *v, *p_proj;

 v = gsl_vector_alloc(2);
 p_proj = gsl_vector_alloc(2);

 for( i = hp->first_frame; i <= hp->last_frame; i++ ){
      cameras_apply( c, point, p_proj, i );
      gsl_vector_set( v, 0, gsl_matrix_get( hp->projs, i - hp->first_frame, 0 ));
      gsl_vector_set( v, 1, gsl_matrix_get( hp->projs, i - hp->first_frame, 1 ));
      gsl_vector_sub( v, p_proj );
      error = gsl_blas_dnrm2(v);
      if( error < tol_error ){
         ninliers++;
         hp_set_inlier( hp, i, TRUE );
      }
      else
         hp_set_inlier( hp, i, FALSE );
 }

 gsl_vector_free(v);
 gsl_vector_free(p_proj);
 return ninliers;
}


void adjust_inliers( HomologPoint *hp, Boolean *inliers )
{
 int i;

 for( i = 0; i < (hp->last_frame - hp->first_frame + 1); i++ )
      hp->is_inlier[i] = inliers[i];
}


double hp_reproj_error( HomologPoint *hp, int i, Cameras *c )
{
 double error = 0;
 gsl_vector *v, *p_proj;

 v = gsl_vector_alloc(2);
 p_proj = gsl_vector_alloc(2);

 cameras_apply( c, hp->p, p_proj, i );
```

```
  gsl_vector_set( v, 0, gsl_matrix_get( hp->projs, i - hp->first_frame, 0 ));
  gsl_vector_set( v, 1, gsl_matrix_get( hp->projs, i - hp->first_frame, 1 ));
  gsl_vector_sub( v, p_proj );
  error = gsl_blas_dnrm2(v);

  gsl_vector_free(v);
  gsl_vector_free(p_proj);

  return error;
}

// relax/relax_cameras.h

#ifndef RELAX_CAMERAS_H
#define RELAX_CAMERAS_H

#include "rcontext.h"
#include "relax_points.h"
#include "ba.h"


void relax_cameras( gsl_vector *xout, RelaxationContext *rc, int cam_id );
void adjust_relax_context( RelaxationContext *rc, gsl_vector *xout, int cam_id );

#endif

// relax/relax_cameras.c

#include "relax_cameras.h"


typedef struct RelaxCamData{
 int cam_id;
 RelaxationContext *rc;
} RelaxCamData;


static void encoder_camera( gsl_vector *xout, gsl_matrix *r, gsl_vector *t );
static void decoder_camera( gsl_matrix *r, gsl_vector *t, gsl_vector *x );
static int relax_cam_cost_func( const gsl_vector *x, void *params, gsl_vector *f);
double relax_cam_reproj_error( gsl_vector *x, gsl_matrix *k, gsl_matrix *r,
                        gsl_vector *t, int cam_id, int point_id,
                        RelaxationContext *rc );


void relax_cameras( gsl_vector *xout, RelaxationContext *rc, int cam_id )
{
 int i, nprojs = 0;
 const gsl_multifit_fdfsolver_type *t = gsl_multifit_fdfsolver_lmsder;
 gsl_multifit_fdfsolver *s;
 gsl_multifit_function_fdf f;
 RelaxCamData d;
 gsl_vector *x;

 x = gsl_vector_alloc(6);
 encoder_camera( x, rc->c->r[cam_id], rc->c->t[cam_id] );

 d.cam_id = cam_id;
 d.rc = rc;

 for( i = 0; i < rc->npoints; i++ )
   if( hp_is_inlier( rc->hp[i], cam_id ) )
     nprojs++;
```

```
 f.f = &relax_cam_cost_func;
 f.df = NULL;

 f.p = 6;
 f.n = nprojs;
 if( f.n < f.p ){
   gsl_vector_memcpy( xout, x );
   return;
 }
 f.params = &d;

 s = gsl_multifit_fdfsolver_alloc(t, f.n, f.p);
 gsl_multifit_fdfsolver_set(s, &f, x);
 relax_optimize(s);
 gsl_vector_memcpy( xout, s->x );
 gsl_multifit_fdfsolver_free(s);
 gsl_vector_free(x);
}


int relax_cam_cost_func(const gsl_vector *x, void *params, gsl_vector *f)
{
 int i, j = 0;
 RelaxCamData *data = (RelaxCamData*)params;
 gsl_matrix *r;
 gsl_vector *t;

 r = gsl_matrix_alloc( 3, 3 );
 t = gsl_vector_alloc( 3 );

 decoder_camera( r, t, x );

 for( i = 0; i < data->rc->npoints; i++ )
   if( hp_is_inlier( data->rc->hp[i], data->cam_id ) ){
     gsl_vector_set( f, j, relax_cam_reproj_error( data->rc->hp[i]->p,
                     data->rc->c->k , r, t, data->cam_id, i, data->rc ));
     j++;
   }

 gsl_matrix_free( r );
 gsl_vector_free( t );
 return GSL_SUCCESS;
}


double relax_cam_reproj_error( gsl_vector *x, gsl_matrix *k, gsl_matrix *r,
                               gsl_vector *t,
                               int cam_id, int point_id, RelaxationContext *rc )
{
 double error;
 gsl_vector *proj, *ref_proj;
 gsl_matrix *p;

 proj = gsl_vector_alloc(2);
 ref_proj = gsl_vector_alloc(2);
 p = gsl_matrix_alloc( 3, 4 );
 HomologPoint *hp;

 hp = rc->hp[point_id];
 calib_pmatrix_make( p, k, r, t );
 calib_apply_P( p, x, proj );
 gsl_vector_set( ref_proj, 0, gsl_matrix_get( hp->projs,
                 cam_id - hp->first_frame, 0 ));
 gsl_vector_set( ref_proj, 1, gsl_matrix_get( hp->projs,
```

```
                cam_id - hp->first_frame, 1 ));
 gsl_vector_sub( proj, ref_proj );
 error = gsl_blas_dnrm2( proj );

 gsl_vector_free(ref_proj);
 gsl_vector_free(proj);
 gsl_matrix_free(p);
 return error;
}


void decoder_camera( gsl_matrix *r, gsl_vector *t, gsl_vector *x )
{
 gsl_vector *aux;

 aux = gsl_vector_alloc(3);
 gsl_vector_set( aux, 0, gsl_vector_get( x, 0 ) );
 gsl_vector_set( aux, 1, gsl_vector_get( x, 1 ) );
 gsl_vector_set( aux, 2, gsl_vector_get( x, 2 ) );
 ba_axis_angle_to_r( r, aux );
 gsl_vector_set( t, 0, gsl_vector_get( x, 3 ) );
 gsl_vector_set( t, 1, gsl_vector_get( x, 4 ) );
 gsl_vector_set( t, 2, gsl_vector_get( x, 5 ) );

 gsl_vector_free( aux );
}


void encoder_camera( gsl_vector *xout, gsl_matrix *r, gsl_vector *t )
{
 gsl_vector *axis_angle;

 axis_angle = gsl_vector_alloc(3);
 ba_r_to_axis_angle( axis_angle, r );
 gsl_vector_set( xout, 0, gsl_vector_get( axis_angle, 0 ) );
 gsl_vector_set( xout, 1, gsl_vector_get( axis_angle, 1 ) );
 gsl_vector_set( xout, 2, gsl_vector_get( axis_angle, 2 ) );
 gsl_vector_set( xout, 3, gsl_vector_get( t, 0 ) );
 gsl_vector_set( xout, 4, gsl_vector_get( t, 1 ) );
 gsl_vector_set( xout, 5, gsl_vector_get( t, 2 ) );

 gsl_vector_free( axis_angle );
}


void adjust_relax_context( RelaxationContext *rc, gsl_vector *xout, int cam_id )
{
 decoder_camera( rc->c->r[cam_id], rc->c->t[cam_id], xout );
}


// relax/relax_points.h

#ifndef RELAX_POINTS_H
#define RELAX_POINTS_H

#include "homol_point.h"
#include "cameras.h"


void relax_points( gsl_vector *xout, gsl_vector *x, HomologPoint *hp, Cameras *c );
void relax_optimize( gsl_multifit_fdfsolver *s );
```

```
#endif

// relax/relax_points.c

#include "rcontext.h"
#include "relax_points.h"


typedef struct RelaxData{
 HomologPoint *hp;
 Cameras *c;
} RelaxData;

static int relax_cost_func( const gsl_vector *x, void *params, gsl_vector *f);
static double relax_reproj_error( const gsl_vector *x, int i, RelaxData *d );


void relax_points( gsl_vector *xout, gsl_vector *x, HomologPoint *hp, Cameras *c )
{
 const gsl_multifit_fdfsolver_type *t = gsl_multifit_fdfsolver_lmsder;
 gsl_multifit_fdfsolver *s;
 gsl_multifit_function_fdf f;
 RelaxData d;

 d.hp = hp;
 d.c = c;

 f.f = &relax_cost_func;
 f.df = NULL;
 f.p = 3;
 f.n = hp->last_frame - hp->first_frame + 1;
 if( f.n < f.p ){
   gsl_vector_memcpy( xout, x );
   return;
 }

 f.params = &d;

 s = gsl_multifit_fdfsolver_alloc(t, f.n, f.p);
 gsl_multifit_fdfsolver_set(s, &f, x);
 relax_optimize(s);
 gsl_vector_memcpy( xout, s->x );
 gsl_multifit_fdfsolver_free(s);
}


int relax_cost_func(const gsl_vector *x, void *params, gsl_vector *f)
{
 int i;
 RelaxData *data = (RelaxData*)params;

 gsl_vector_set_zero(f);
 for( i = data->hp->first_frame; i <= data->hp->last_frame; i++ )
   if( hp_is_inlier( data->hp, i ) ){
     gsl_vector_set( f, i - data->hp->first_frame,
                     relax_reproj_error( x, i, data ));
   }

 return GSL_SUCCESS;
}


double relax_reproj_error( const gsl_vector *x, int i, RelaxData *data )
{
```

```
  double error;
  gsl_vector *proj, *ref_proj;

  proj = gsl_vector_alloc(2);
  ref_proj = gsl_vector_alloc(2);

  cameras_apply( data->c, x, proj, i );

  gsl_vector_set( ref_proj, 0, gsl_matrix_get( data->hp->projs,
                  i - data->hp->first_frame, 0 ));
  gsl_vector_set( ref_proj, 1, gsl_matrix_get( data->hp->projs,
                  i - data->hp->first_frame, 1 ));
  gsl_vector_sub( proj, ref_proj );
  error = gsl_blas_dnrm2( proj );

  gsl_vector_free(proj);
  gsl_vector_free(ref_proj);
  return error;
}


void relax_optimize( gsl_multifit_fdfsolver *s )
{
  int status, iter = 0;

  do{
     iter++;
     status = gsl_multifit_fdfsolver_iterate(s);
     if(status)
        break;
     status = gsl_multifit_test_delta(s->dx, s->x, LM_EPS, LM_EPS);
     printf( "%i Error = %f\n", iter, gsl_blas_dnrm2(s->f) );
  }
  while( (status == GSL_CONTINUE) && (iter < LM_MAX_ITERATIONS ) );
}


// relax/rcontext.h

#ifndef REL_CONTEXT_H
#define REL_CONTEXT_H

#include "homol_point.h"
#include "cameras.h"
#include "ba.h"
#include "featureslist.h"

#define MAX_POINTS 2000
#define MAX_FEATURES 1000
#define MIN_HOMOL_SEGMENT 45
#define MIN_HOMOL_INLIERS 45

typedef struct RelaxationContext{
   HomologPoint *hp[MAX_POINTS];
   int npoints;
   Cameras *c;
   gsl_matrix *xproj;
   gsl_matrix *yproj;
   int last_features[MAX_FEATURES];
} RelaxationContext;


RelaxationContext *rc_alloc( char *fname, Cameras *c, double tol_error );
void  rc_set_all_projs( RelaxationContext *rc, int j, int feature );
```

```
void rc_free( RelaxationContext *rc );

#endif

// relax/rcontext.c

#include "rcontext.h"

static int get_nframes( char *fname );


int get_nframes( char *fname )
{
 FILE *fin;
 int i, nfeatures, status, frame_id, nframes = 0;
 double coordx, coordy;
 fin = fopen( fname, "r" );

 fscanf( fin, "Features per frame = %i\n", &nfeatures );
 while( !feof(fin) ){
     for( i = 0; i < nfeatures; i++ ){
       fscanf( fin, "Frame %i\n", &frame_id );
       fscanf( fin, "%lf %lf %i\n", &coordx, &coordy, &status );
     }
     nframes++;
 }
 fclose(fin);
 return nframes;
}


#define NOT_TRACKED -1

 RelaxationContext *rc_alloc( char *fname, Cameras *c, double tol_error )
{
 FILE *fin;
 int i, j=0, k=0, nfeatures, nframes, status,
     frame_id=0, last_frame = 0;
 double coordx, coordy;
 RelaxationContext *rc;

 rc = NEWSTRUCT( RelaxationContext);
 nframes = get_nframes( fname );
 fin = fopen( fname, "r" );
 fscanf( fin, "Features per frame = %i\n", &nfeatures );
 rc->xproj = gsl_matrix_alloc( nframes, nfeatures );
 rc->yproj = gsl_matrix_alloc( nframes, nfeatures );

 for( i = 0; i < MAX_FEATURES; i++ )
   rc->last_features[i] = NOT_TRACKED;

 while( !feof(fin) && (frame_id <= c->ncams) ) {
     for( i = 0; i < nfeatures; i++ ){
       fscanf( fin, "Frame %i\n", &frame_id );
       fscanf( fin, "%lf %lf %i\n", &coordx, &coordy, &status );
       gsl_matrix_set( rc->xproj, frame_id, i, coordx );
       gsl_matrix_set( rc->yproj, frame_id, i, coordy );

       if( rc->last_features[i] == NOT_TRACKED )
           rc->last_features[i] =  frame_id;

       if( ( frame_id - rc->last_features[i] > MIN_HOMOL_SEGMENT ) &&
           (( frame_id == c->ncams ) ||  (status != FEATURE_TRACKED )) ){
```

```
            rc->hp[j] = hp_alloc( rc->last_features[i], frame_id - 1 );
            rc->last_features[i] = frame_id - 1;
            rc_set_all_projs( rc, j, i );
            hp_reconstruct_points( rc->hp[j], c, tol_error );

            if( rc->hp[j]->ninliers > MIN_HOMOL_INLIERS ){
              printf( "%i: [%i,%i]: %i ninliers: %i\n", k++, rc->hp[j]->first_frame,
                       rc->hp[j]->last_frame,
                       rc->hp[j]->last_frame - rc->hp[j]->first_frame + 1,
                       rc->hp[j]->ninliers );
              j++;
            }
         }
      }
 }

 rc->c = c;
 rc->npoints = j;
 return rc;
}


void  rc_set_all_projs( RelaxationContext *rc, int j, int feature )
{
 int i;

 for( i = rc->hp[j]->first_frame; i <= rc->hp[j]->last_frame; i++ ){
      hp_set_proj( rc->hp[j], i,
                   gsl_matrix_get( rc->xproj, i, feature ),
                   gsl_matrix_get( rc->yproj, i, feature ) ) ;

 }
}


void rc_free( RelaxationContext *rc )
{
 int i;

 for( i = 0; i < rc->npoints; i++ )
   hp_free( rc->hp[i] );

 cameras_free( rc->c );
}
```

## 7.16 MATCHMOVE PROGRAM

This is the MatchMove program. It estimates a first set of extrinsic parameters related to each frame, that are after refined by the Relaxation Program.

argv[1] receives the name of the file that contains the intrinsic parameter matrix.

argv[2] receives the name of the output file generated by the KLT Program, presented in the Chapter 6.

argv[3] receives the name of the output file.

```
// calib/main.c

#include "sequence.h"
```

```
static void write_cameras( FILE *f, CalibSequenceContext *cc );

int main( int argc, char **argv )
{
 int i;
 FILE *fin, *fout, *fk;
 CalibSequenceContext *cc;
 SubSeqNode *aux;
 gsl_matrix *k;
 FeaturesList *fl;

 init_timer_random();
 k = gsl_matrix_alloc( 3, 3 );
 fk = fopen( argv[1], "r" );
 gsl_matrix_fscanf( fk, k );
 fin = fopen( argv[2], "r" );
 fout = fopen( argv[3], "w" );
 cc = calib_context_alloc( features_list_read( fin ) );
 seq_extrinsic_calibrate( cc, k );

 write_cameras( fout, cc );
 calib_context_dispose( cc );
 gsl_matrix_free( k );
 fclose( fin );
 fclose( fout );
 fclose( fk );
}


void write_cameras( FILE *f, CalibSequenceContext *cc )
{
 int i;
 SubSeqNode *aux;

 for(  aux = cc->sl; (aux != NULL) && (aux->next != NULL); aux = aux->next ){
   for( i = 0; i<aux->nframes - 1 ; i++ ){
     fprintf( f, "Frame %i\n", aux->first_frame + i );
     gsl_matrix_fprintf( f, aux->cl[i].r, "%f" );
     fprintf( f, "\n" );
     gsl_vector_fprintf( f, aux->cl[i].t, "%f" );
     fprintf( f, "\n" );
   }
 }
}
```

## 7.17 RELAXATION PROGRAM

This program uses the Relaxation API to perform the relaxation cycles. It improves the quality of the calibration made by the program defined in the Section 7.16.

argv[1] receives the name of the file that contains the intrinsic parameter matrix used for capturing the video.

argv[2] receives the name of the output file generated by the KLT Program, presented in the Chapter 6.

argv[3] receives the name of the file that contains the output of the MatchMove Program.

argv[4] receives the name of the output file.

```
// relax/main.c

#include "rcontext.h"
#include "ba.h"
#include "relax_points.h"
#include "relax_cameras.h"

#define NRELAX 9
#define NCYCLES 10    /* Number of complete relax until changing tol */
#define FIRST_RANSAC_TOL 10.
#define TOL_STEP 1.

static void relax( RelaxationContext *rc, Cameras *cam, int niters,
            double first_tol, double tol_step );
static void optimize( RelaxationContext *rc, Cameras *cam );
static void refine_inliers( HomologPoint *hp, Cameras *c , double tol );

void main( int argc, char **argv )
{
 RelaxationContext *rc;
 HomologPoint *hp;
 Cameras *cam;
 int i, nframes;
 gsl_matrix *k;
 int m[200];

 k = gsl_matrix_alloc(3,3);

 FILE *kf = fopen( argv[1], "r" );
 FILE *cam_f = fopen( argv[3], "r" );
 FILE *fout = fopen( argv[4], "w" );

 gsl_matrix_fscanf( kf, k );
 cam = cameras_read( k, cam_f );
 rc = rc_alloc( argv[2], cam, 10. );

 relax( rc, cam, NRELAX, FIRST_RANSAC_TOL, TOL_STEP );

 for( i = 0; i < cam->ncams; i++ ){
   fprintf( fout, "Frame %i\n", i );
   gsl_matrix_fprintf( fout, rc->c->r[i], "%f" );
   fprintf( fout, "\n" );
   gsl_vector_fprintf( fout, rc->c->t[i], "%f" );
   fprintf( fout, "\n" );
 }

 fclose( fout );
 fclose( kf );
 fclose( cam_f );
 gsl_matrix_free(k);
}

void relax( RelaxationContext *rc, Cameras *cam, int niters,
            double first_tol, double tol_step )
{
  int i, j, total_ninliers = 0;
  double tol = first_tol;

  for( j = 0; j < niters; j++ ){
     for( i = 0; i < NCYCLES; i++ ){
        optimize( rc, cam );
     }
     printf( "MAX REPROJ ERROR: %lf\n", tol );
```

```
    for( i=0; i < rc->npoints; i++ ){
      refine_inliers( rc->hp[i], cam, tol );
      printf( "%i: [%i,%i], %i ninliers = %i\n", i, rc->hp[i]->first_frame,
                       rc->hp[i]->last_frame,
                       rc->hp[i]->last_frame - rc->hp[i]->first_frame + 1,
                       rc->hp[i]->ninliers );
      total_ninliers += rc->hp[i]->ninliers;
    }
    printf( "Total NINLIERS = %i\n", total_ninliers );
    tol -= tol_step;
  }
}

void optimize( RelaxationContext *rc, Cameras *cam )
{
 int i;
 gsl_vector *xpts, *xcams;

 xpts= gsl_vector_alloc(3);
 xcams = gsl_vector_alloc(6);

 for( i = 0; i < rc->npoints; i++ ){
    printf( "point %i \n", i  );
    relax_points( xpts, rc->hp[i]->p, rc->hp[i], cam );
    gsl_vector_memcpy( rc->hp[i]->p, xpts );
 }

 for( i=0; i < rc->c->ncams; i++ ){
    printf( "camera %i \n", i  );
    relax_cameras( xcams, rc, i );
    adjust_relax_context( rc, xcams, i );
 }

 gsl_vector_free(xpts);
 gsl_vector_free(xcams);
}

void refine_inliers( HomologPoint *hp, Cameras *c , double tol )
{
 int i;
 hp->ninliers = 0;

 for( i = hp->first_frame; i <= hp->last_frame; i++ )
   if( hp_reproj_error( hp, i, c ) < tol ){
      hp_set_inlier( hp, i, TRUE );
      hp->ninliers++;
   }
   else
      hp_set_inlier( hp, i, FALSE );
}
```

CHAPTER 8

# Modeling Tools

An important tool that we need for the creation of a visual effect is a software able to measure the coordinates of points in the scene reference used by the programs defined in the Section 7.17. These 3D coordinates can be processed by other Modeling tools created by the reader to correctly positioning objects in the scene.

This tool consists of two parts, the first one is a program that is used to define the projections of a set of 3D points over many frames. This program is a simple 2D graphical interface program created using the OpenCV Library. The output of it is a list of 2D projections of points over all selected frames by the user.

The second part of the tool processes the output of the first part. It generates the 3D coordinates of $n$ points in the form of an $n \times 3$ matrix represented by a outputted file by the Gnu Scientific Library. This program firstly finds an initial estimation of the 3D coordinates of the points using the algorithm presented in the Section 5.8.2, choosing the best estimation made using pair of frames. After that, the coordinates are refined by a Levenberg-Marquardt algorithm that minimizes the projection error over the frames assuming that the cameras are fixed.

## 8.1 API

```
void pcloud_init( PointCloud *pc, int frame, int npoints );
```

This function initializes the structure *pc* with a 2D point cloud. *frame* specifies the frame in which the points are defined, and *npoints* defines the number of points of the point cloud.

```
void pcloud_free( PointCloud *pc );
```

This function dealocates the 2D point cloud *pc*.

```
void pcloud_get_point( gsl_vector *v, PointCloud *pc,
                       int i );
```

DOI: 10.1201/9781003206026-8

This function gets the $i^{th}$ point of the 2D point cloud *pc* and returns in the vector $v$.

```
void pcloud_set_point( PointCloud *pc, int i,
                       gsl_vector *v );
```

This function sets the $i^{th}$ point of the 2D point cloud *pc* with the coordinates specified by the vector $v$.

```
int pcloud_get_npoints( PointCloud *pc );
```

This function returns the number of 2D points of the point cloud *pc*.

```
void pcloud_read( PointCloud *pc, FILE *f );
```

This function reads a 2D points cloud related to a frame over the file $f$, and returns it in the structure *pc*.

```
gsl_matrix* pcloud_calib( PointCloud *pc, Cameras *cam,
                          int nframes );
```

This function returns a $n \times 3$ matrix representing the 3D vectors correspondent to the reconstruction of the array of 2D point clouds *pc*, over $nframes$ frames. *cam* is a structure estimated by the Relaxation module (Section 7.15.10) that contains the intrinsic and extrinsic parameters of the cameras related to a video.

```
void pcloud_nlin_optimize( gsl_vector *xout, gsl_vector *x,
                           Cameras *cams, PointCloud *pc,
                           int nframes, int point );
```

This is an auxiliary function responsible to refine the reconstruction of each 3D point. $x$ is the initial estimation of the 3D coordinate of the point, $xout$ is the final estimation calculated by the Levenberg-Marquardt algorithm. *pc* is a pointer to a array of 2D point cloud structures. $nframes$ is the number of frames in *pc*, and *point* is the index of the correspondent point that is being optimized.

## 8.2 CODE

```
// pcloud/pcloud.h

#ifndef POINT_CLOUD_H
#define POINT_CLOUD_H
```

```
#include "ba.h"
#include "defs.h"
#include "calib.h"
#include "cameras.h"


typedef struct PointCloud{
  int frame;
  gsl_matrix *m;
} PointCloud;


/* Basic functions */
void pcloud_init( PointCloud *pc, int frame, int npoints );
void pcloud_free( PointCloud *pc );
void pcloud_get_point( gsl_vector *v, PointCloud *pc, int i );
void pcloud_set_point( PointCloud *pc, int i,  gsl_vector *v );
int pcloud_get_npoints( PointCloud *pc );
void pcloud_read( PointCloud *pc, FILE *f );

/* 3D Reconstruct Functions */
gsl_matrix* pcloud_calib( PointCloud *pc, Cameras *cam, int nframes );

/* Reconstruction using Non Linear Optimization */
void pcloud_nlin_optimize( gsl_vector *xout, gsl_vector *x, Cameras *cams,
                           PointCloud *pc, int nframes, int point );

#endif


// pcloud/pcloud.c

#include "pcloud.h"


void pcloud_init( PointCloud *pc, int frame, int npoints )
{
 pc->m = gsl_matrix_alloc( npoints, 2 );
 pc->frame = frame;
}


void pcloud_free( PointCloud *pc )
{
 gsl_matrix_free( pc->m );
 free( pc );
}


void pcloud_get_point( gsl_vector *v, PointCloud *pc, int i )
{
  gsl_matrix_get_row( v, pc->m, i );
}


void pcloud_set_point( PointCloud *pc, int i,  gsl_vector *v )
{
  gsl_matrix_set_row( pc->m, i, v );
}


int pcloud_get_npoints( PointCloud *pc )
{
 return pc->m->size1;
```

```
}


void pcloud_read( PointCloud *pc, FILE *f )
{
 double val;
 int i, j, frame, npoints;
 fscanf( f, "Frame %i\n", &frame );
 fscanf( f, "Npoints %i\n", &npoints );
 pcloud_init( pc, frame, npoints );

 for( i = 0; i < npoints; i++ )
    for( j = 0; j < 2; j++ ){
      fscanf( f, "%lf", &val );
      gsl_matrix_set( pc->m, i, j, val );
    }
 fscanf( f, "\n" );
}


// pcloud/pccalib.c

#include "pcloud.h"


static void reconstruct_point( gsl_vector *xout, PointCloud *pc,
                               Cameras *cam, int nframes, int point );
static Real reproj_error( gsl_vector *x,  Cameras *cams, PointCloud *pc,
                          int point, int nframes );


gsl_matrix* pcloud_calib( PointCloud *pc, Cameras *cam, int nframes )
{
 gsl_vector *x, *xout;
 gsl_matrix *output;
 int point, npoints;

 npoints = pcloud_get_npoints(pc);

 output = gsl_matrix_alloc( npoints, 3 );
 x = gsl_vector_alloc( 3 );
 xout = gsl_vector_alloc( 3 );

 for( point = 0; point < npoints; point++ ){
    reconstruct_point( x, pc, cam, nframes, point );
    pcloud_nlin_optimize( xout, x, cam, pc, nframes, point );
    gsl_matrix_set_row( output, point, xout );
 }

 gsl_vector_free(x);
 gsl_vector_free(xout);

 return output;
}


void reconstruct_point( gsl_vector *xout, PointCloud *pc, Cameras *cam,
                        int nframes, int point )
{
 int i, j;
 double error, min_reproj_error = MAX_FLOAT;
 gsl_vector *v1, *v2, *x;
 gsl_matrix *p1, *p2;
```

```
 p1 = gsl_matrix_alloc( 3, 4 );
 p2 = gsl_matrix_alloc( 3, 4 );
 v1 = gsl_vector_alloc( 2 );
 v2 = gsl_vector_alloc( 2 );
 x = gsl_vector_alloc( 3 );

 for( i = 0; i < nframes; i++ )
   for( j = i+1; j < nframes; j++ ){
      calib_pmatrix_make( p1, cam->k, cam->r[pc[i].frame], cam->t[pc[i].frame] );
      calib_pmatrix_make( p2, cam->k, cam->r[pc[j].frame], cam->t[pc[j].frame] );
      pcloud_get_point( v1, &pc[i], point );
      pcloud_get_point( v2, &pc[j], point );
      calib_get_3dpoint( p1, p2, v1, v2, x );
      if( (error = reproj_error( x, cam, pc, point, nframes )) < min_reproj_error ){
         min_reproj_error = error;
         gsl_vector_memcpy( xout, x );
      }
   }

  gsl_matrix_free( p1 );
  gsl_matrix_free( p2 );
  gsl_vector_free( v1 );
  gsl_vector_free( v2 );
  gsl_vector_free( x );
}


Real reproj_error( gsl_vector *x,  Cameras *cams, PointCloud *pc,
                   int point, int nframes )
{
 int j;
 Real error = 0;
 gsl_vector *x_proj, *p;

 x_proj = gsl_vector_alloc(2);
 p = gsl_vector_alloc(2);

 for( j = 0; j < nframes; j++ )
   if( pc[j].frame < cams->ncams ){
      cameras_apply( cams, x, x_proj, pc[j].frame );
      pcloud_get_point( p, &pc[j], point );
      gsl_vector_sub( x_proj, p );
      error += gsl_blas_dnrm2( x_proj );
   }

 gsl_vector_free( x_proj );
 gsl_vector_free( p );
 return error;
}


// pcloud/pcnlin.c

#include "pcloud.h"


typedef struct PCloudData{
  Cameras *cams;
  PointCloud *pc;
  int nframes;
  int point;
} PCloudData;
```

```
static int cost_func(const gsl_vector *x, void *params, gsl_vector *f );
static float reproj_point_error( gsl_vector *x, PCloudData *data , int frame );
static void pcnlin_optimize( gsl_multifit_fdfsolver *s  );


void pcloud_nlin_optimize( gsl_vector *xout, gsl_vector *x, Cameras *cams,
                        PointCloud *pc, int nframes, int point )
{
 const gsl_multifit_fdfsolver_type *t = gsl_multifit_fdfsolver_lmsder;
 gsl_multifit_fdfsolver *s;
 gsl_multifit_function_fdf f;
 PCloudData d;

 f.f = &cost_func;
 f.df = NULL;
 f.p = 3;
 f.n = nframes;

 d.cams = cams;
 d.pc = pc;
 d.nframes = nframes;
 d.point = point;

 f.params = &d;

 s = gsl_multifit_fdfsolver_alloc(t, f.n, f.p);
 gsl_multifit_fdfsolver_set(s, &f, x);

 pcnlin_optimize(s);
 gsl_vector_memcpy( xout, s->x );
 gsl_multifit_fdfsolver_free(s);
}


int cost_func(const gsl_vector *x, void *params, gsl_vector *f)
{
 int i;
 PCloudData *data = (PCloudData*)params;

 gsl_vector_set_zero(f);
 for( i = 0; i < data->nframes; i++ )
     gsl_vector_set( f, i, reproj_point_error( x, data, i ) );

 return GSL_SUCCESS;
}


float reproj_point_error( gsl_vector *x, PCloudData *data, int frame )
{
 float error;
 gsl_vector *p, *x_proj;

 p = gsl_vector_alloc(2);
 x_proj = gsl_vector_alloc(2);

 cameras_apply( data->cams, x, x_proj, data->pc[frame].frame );
 pcloud_get_point( p, &(data->pc[frame]), data->point );
 gsl_vector_sub( x_proj, p );
 error = gsl_blas_dnrm2( x_proj );

 gsl_vector_free( p );
 gsl_vector_free( x_proj );
 return error;
}
```

```
void pcnlin_optimize( gsl_multifit_fdfsolver *s )
{
 int status, iter = 0;

 do{
    iter++;
    status = gsl_multifit_fdfsolver_iterate(s);
    if(status)
       break;
    status = gsl_multifit_test_delta(s->dx, s->x, LM_EPS, LM_EPS);
 }
 while( (status == GSL_CONTINUE) && (iter < LM_MAX_ITERATIONS ) );
}
```

## 8.3 POINT CLOUD DEFINER PROGRAM

This program is a graphical interface, built using the OpenCV Library, allowing the user to define a 3D point cloud.

argv[1] receives the name of the video file.

argv[2] receives the number of 3D points that the user wants to capture.

argv[3] receives the name of the file that contains the extrinsic parameter of the cameras. This file is generated by the programs defined in the Sections 7.16 or 7.17.

argv[4] receives a factor of scale applied to the frames of the video. It can be used to enlarge or reduces the size of the image presented in the interface.

The appearance of the interface is presented in Figure 8.1. There is a slider that can be used to change the frame number. The user must select the frame number and mark points on then. It can be done by pointing the mouse over some point over the frame, and clicking the left button of the mouse. Sequentially, the graphical interface will number the marked point.

Figure 8.1 Graphical Interface of the Definer Point Cloud Program.

After selecting the correct number of points in the frame, the user must press the space bar.

The user is responsible to select the correct quantity of points specified in the argv[2] parameter, and must choose them in the same order in all the selected frames.

The output of the program is written in *stdout*. The user can redirect it to a file using the ">" Unix-like operator or can be passed to the program *pclcalib* defined in Section 8.4 using the "|" operator.

```
// defpcl/main.cxx

#include "opencv2/imgcodecs.hpp"
#include "opencv2/highgui.hpp"
#include <opencv2/imgproc.hpp>
#include <iostream>
#include <vector>

using namespace std;
using namespace cv;

float scale;
Mat image;
int nframes;
int npoints;
int frame_id;
vector<Mat> video;
Point p[20];

static void load_video( vector<Mat> &video, char *fname );
static void on_trackbar( int, void* );
static void CallBackFunc(int event, int x, int y, int flags, void* userdata);
static int get_ncameras( FILE *f );


int main( int argc, char **argv )
{
   scale = atof( argv[4] );
   load_video( video, argv[1] );
   npoints = atoi( argv[2] );
   FILE *cam_file = fopen( argv[3], "r" );
   int nframes = get_ncameras( cam_file );
   fclose( cam_file );
   frame_id = 0;

   imshow( "PCloud", video[frame_id] );
   char TrackbarName[50];
   sprintf( TrackbarName, "%d Frames", nframes );
   createTrackbar( TrackbarName, "PCloud", &frame_id, nframes-1, on_trackbar );
   on_trackbar( frame_id, 0 );
   setMouseCallback("PCloud", CallBackFunc, NULL);

   for(;;){
       imshow( "PCloud", video[frame_id] );
       char c = waitKey();
       if( c == ' ' ){
         cout << "Frame " << frame_id << endl;
         cout << "Npoints " << npoints << endl;
         for( int i = 0; i < npoints; i++ )
            cout << p[i].x/scale << " " << p[i].y/scale << endl;
         cout << endl;
       }
```

```
        else if(c == 27 )
          exit(0);
        else if( getWindowProperty( "PCloud", WND_PROP_VISIBLE) == -1 )
            exit(0);
    }
    return 0;
}


void load_video( vector<Mat> &video, char *fname )
{
 Mat frame;
 Mat  *resized;
 VideoCapture capture(fname);
 do{
      resized = new Mat();
      capture >> frame;
      if( !frame.empty() ){
         resize(frame, *resized, Size(scale*frame.cols, scale*frame.rows),
                                             INTER_LINEAR);
         video.push_back(*resized);
      }
 }while(!frame.empty());
}


void on_trackbar( int, void* )
{
  imshow( "PCloud", video[frame_id] );
}


void CallBackFunc(int event, int x, int y, int flags, void* userdata)
{
 static int i = 0;
 char str[2];

 if( i == 0 )
    image = video[frame_id].clone();

 if( event == EVENT_LBUTTONDOWN ){
    imshow( "PCloud", video[frame_id] );
    line( image, Point( x - 5, y ), Point( x + 5, y ), Scalar( 0, 0, 255 ),  2, 8 );
    line( image, Point( x, y - 5 ), Point( x, y + 5 ), Scalar( 0, 0, 255 ),  2, 8 );
    sprintf( str, "%i", i+1 );
    putText( image, str, Point( x-15, y -10), FONT_HERSHEY_DUPLEX, 1,
    Scalar(0,0,255), 2 );
    imshow( "PCloud", image );
    p[i].x = x;
    p[i].y = y;
    i++;
 }

 if( i == npoints )
   i = 0;
}


int get_ncameras( FILE *f )
{
 int i = 0;
 float r, t;

 while( !feof(f) ){
```

```
    fscanf( f, "Frame %i\n", &i );
    fscanf( f, "%f%f%f%f%f%f%f%f%f\n",
                &r, &r, &r,  &r, &r, &r, &r, &r, &r );
    fscanf( f, "%f%f%f\n", &t, &t, &t );
    i++;
  }
  return i;
}
```

## 8.4 POINT CLOUD CALIB PROGRAM

This program processes the output of the program presented on the Section 8.3.

argv[1] receives the name of the file that contains the matrix of intrinsic parameters used in the video capture. It can be generated, for example, using the program defined in Section 4.11.

argv[2] receives the name of the file that contains the matrix of extrinsic parameters used in the video capture. This file is must be previously generated using the program presented on Section 7.17.

*stdin* receives the output of the program defined on Section 8.3.

The output is written in *stdout*, and can be redirected to others modeling programs created by the user.

```
// pclcalib/main.c

#include "pcloud.h"
#include "cameras.h"

#define MAX_NCLOUDS 20

static void write_cloud( gsl_matrix *m );

int main( int argc, char **argv )
{
 int i = 0;
 FILE *k_file, *cam_file;
 Cameras *cams;
 PointCloud pc[MAX_NCLOUDS];
 gsl_matrix *m;
 gsl_matrix *k;

 k = gsl_matrix_alloc( 3, 3 );
 k_file = fopen( argv[1], "r" );
 gsl_matrix_fscanf( k_file, k );
 cam_file = fopen( argv[2], "r" );
 cams = cameras_read( k, cam_file );

 while( !feof(stdin) )
   pcloud_read( &pc[i++], stdin );

 m = pcloud_calib( pc, cams, i );

 gsl_matrix_fprintf( stdout, m, "%f" );

 fclose( k_file );
 fclose( cam_file );
 gsl_matrix_free( k );
```

```
 gsl_matrix_free( m );
}
```

An example of modeling tool that can be used to processes the output of this program is presented bellow. It generates the text of a triangle in the S3D scene language.

```
#include <stdio.h>

void main( int argc, char **argv )
{
 int i, j;
 double m[3][3];

 for( i = 0; i < 3; i++ )
   for( j = 0; j < 3; j++ )
     scanf( "%lf", &m[i][j] );

 printf( "shape = trilist { {{%lf, %lf, %lf},  {%lf, %lf, %lf}, {%lf, %lf, %lf}} }",
          m[0][0], m[0][1], m[0][2],
          m[1][0], m[1][1], m[1][2],
          m[2][0], m[2][1], m[2][2] );
}
```

CHAPTER 9

# Light Transport and Monte Carlo

IN THIS CHAPTER we will present the facts about the light transport that are necessary for creating a high quality rendering, good enough for being used in visual effects.

Firstly it is important to understand that the problem that we are interested is how the energy on a surface in the scene travels to other surface and how this process creates the steady state on the illumination in the scene.

The mathematical object more important to model this problem is the Radiance.

## 9.1 RADIANCE

Radiance is a mathematical object that describes how much the energy incident from some direction increases the incoming intensity of the light over some point on a surface. More precisely, it describes the light incoming intensity over a projected area on a surface.

The physical unit used for describing radiance is $Wm^{-2}sr^{-1}$. It means that if we integrate the light potency coming from all the hemisphere over a point we find the power per square meter over the surface. It means that, in order to compute the potency that a surface receives, it is necessary to integrate over the directions over the hemisphere above the point, and also integrate the result over the surface area (Figure 9.1).

Another important fact is that the light incoming more aligned to the vertical direction of the surface is more relevant for the radiance than the ones incoming in an oblique way. It occurs because the light incoming obliquely is spread over a larger area.

It is not difficult to show that the radiance is proportional to $cos\theta$, in which $\theta$ is the angle between the normal of the surface and the light incidence direction. It happens because if we have a vertical incident light flux arriving over a surface of area $A$ and we make it oblique, making an angle of $\theta$ with the normal, then the surface area becomes to $A/cos\theta$ (Figure 9.2). Consequently, the incoming light per unity of projected surface area reduces when we increases $\theta$.

DOI: 10.1201/9781003206026-9

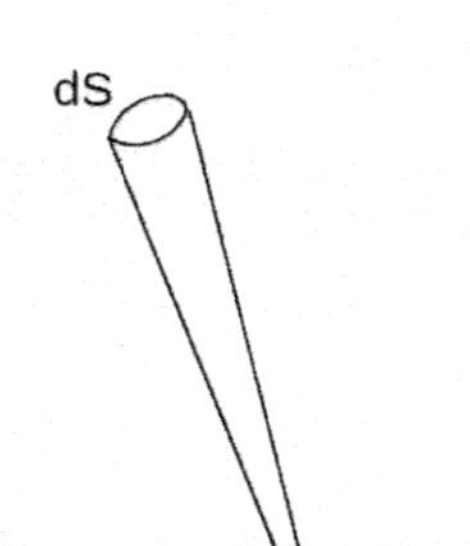


Figure 9.1 Element of integration for an area element. For each surface area element we need to integrate over the hemisphere above the surface.

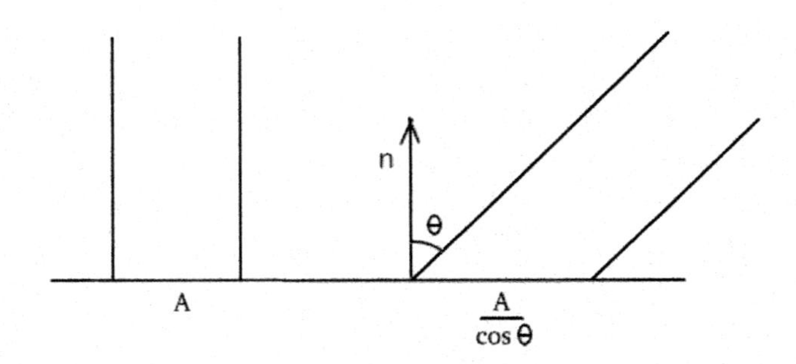


Figure 9.2 On the left, an orthogonal flux of light with section A. On the right, an oblique flux of light with section A. On the right, the light spreads over a larger surface area.

We also have to consider that only the directions on a hemisphere touching the point must be considered.

Join all the previous information; we can conclude that the potency absorbed by a surface $M$ is defined by:

$$I = \int_M \int_{S_x} L(x, u) cos\theta_u \, du \, dx,$$

such that $L(x, u)$ is the radiance from the direction $u$ in the point $x \in M$, and $S_x$ is the hemisphere over $x$.

Considering $n_x \in \mathbb{R}^3$ the unit vector normal to the surface $M$ in the point $x$, we can rewrite that as:

$$I = \int_M \int_{S_x} L(x, u) \langle u, n_x \rangle \, du \, dx,$$

## 9.2 THE INVARIANCE OF THE RADIANCE

A very important property about the definition of Radiance is that it makes it invariant along a path. It means that the radiance emitted by a surface far away from a point over a surface is always the same, besides the distance that the surface emitting light. The proof of this fact can be found in [21]. But it is intuitive, since when we are far away from a surface emitting light energy, the light potency, reduces quadratically in relation to the distance, but the projected area of the surface emitting light also reduces quadratically, making the energy more concentrated in a small surface over the hemisphere over a point in the surface that is receiving light [26].

## 9.3 THE BRDF AND THE RENDERING EQUATION

The Rendering Equation expresses the flow of energy between many surfaces, or in other words, it expresses the Light Transport on a 3D scene. In order to express it, we need a mathematical object able to define how the incident radiance from some direction in the hemisphere over a point is transformed in an output radiance on other direction.

We call this mathematical object BRDF (Bidirectional Reflectance Distribution Function). It's physical unity is $sr^{-1}$, which means that the integration of it over a hemisphere results on a adimensional value.

The BRDF is responsible for combining all the energy coming to a point and find the radiance in an output direction in a linear way.

On other words, if $f(x, u, v)$ is the BRDF, defined in the point $x$, by the incoming direction $u$, and by the output direction $v$, then we have that the radiance on this direction is defined by:

$$L(x, v) = \int_{S_x} f(x, u, v) L(x, u) \langle u, n \rangle \, du$$

This equation expresses the behavior of a reflexive surface on a 3D scene. In order to understand the Light Transport, we must consider that besides reflection, a surface can also inputs energy to the scene (a lamp for example). Thus, the radiance coming from a surface that does both things can be better expressed by:

$$L(x, v) = \int_{S_x} f(x, u, v) L(x, u) \langle u, n \rangle \, du + L_e(x, v), \tag{9.1}$$

such that $L_e(x, v)$ is the output radiance emitted by the surface in the point $x \in M$ to the direction $v$.

The Equation 9.1 connects the points in scene, describing the Light Transport defined by them. It is called in the Computer Graphics area by Rendering Equation.

To generate a 3D image we need to find the steady state defined by it. It is possible to prove that we can guarantee this property if all the BRDF of the surfaces are plausible [17]:

$$\int_{S_x} f(x, u, v) \langle u, n \rangle \, du < 1, \forall v \in S_x$$

which means that any surface in the scene reflects more energy than receives.

The steady state of the Rendering Equation is intuitive, since in our real world experience with the Light Transport, after turning on a lamp in a room, we expect to see a static scene without any change in the brightness of the walls.

## 9.4 OTHER DEFINITION FOR THE RENDERING EQUATION

As we will see, sometimes the Equation 9.1 is not appropriate to be used. On this case, a different, but equivalent version of it is better. This rendering equation is expressed by [21]:

$$L(x, v) = \int_A f(x, u, v) L(x, u) V(x, x') G(x, x')\, dA_{x'} + L_e(x, v), \tag{9.2}$$

in which $x \in A$ is a point in the scene receiving light from the point $x' \in A$, such that $A$ is all the surfaces in the scene.

$V(x, x')$ is defined by 1 if the path between $x$ and $x'$ is free of surfaces in the scene, and its value is 0 otherwise.

$G(x, x')$ is a geometric term defined by:

$$G(x, x') = \frac{\langle u, n \rangle \langle u', n' \rangle}{\|x' - x\|^2}, \tag{9.3}$$

such that $u \in \mathbb{R}^3$ is defined by

$$u = \frac{x' - x}{\|x' - x\|},$$

the vector $n \in S_x$ is the unitary normal of the surface in the point $x$. The vector $u' \in \mathbb{R}^3$ is defined by:

$$u' = \frac{x - x'}{\|x - x'\|},$$

and $n' \in S_{x'}$ is the unitary vector of the surface in the point $x'$ (Figure 9.3).

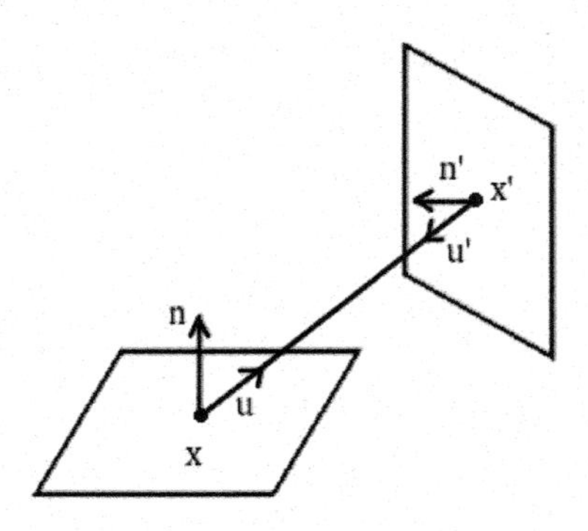


Figure 9.3 Geometric elements defined by the light transport between two points in the scene.

## 9.5 EXAMPLES OF BRDF

In our book we only consider surfaces with three kinds of BRDF: The perfect Lambertian surfaces, the perfect mirror surfaces, and the ones satisfying the Modified Blinn-Phong's model [21].

### 9.5.1 The perfect Lambertian Surface BRDF

A Lambertian surface is a kind of surface that spreads all the incident light over the hemisphere without any preferential direction. It means that the BRDF is constant over the entire hemisphere. Consequently, an observer will see the same result whatever are the direction of the observation. The unique restriction that must be satisfied is that it must be plausible. The behavior of a Lambertian surface is defined by the scattering of the light on the surface after penetrating the boundary of the surface.

### 9.5.2 The Perfect Mirror BRDF

The BRDF of a perfect mirror surface is defined by:

$$f(x, u, v) = \frac{k_s \delta(r - v)}{\langle n, u \rangle}$$

such that $r \in S$ is the perfect mirror reflection in the hemisphere over the point, and $r = u + 2n\langle n, u \rangle$ is the reflection direction [31]. $k_s$ specifies the fraction of the light reflected.

### 9.5.3 The Modified Blinn-Phong's BRDF

The BRDF based on Modified Blinn-Phong's model [21], is made by a linear combination of two parts: a diffuse term, that corresponds to a perfect Lambertian component, and a specular reflection, that can be thought as the light that are reflected by the surface whenever the light finds the boundary of a different refractive index medium. Consequently, it has a preferential direction defined by the normal of the surface.

Being $h(u, v) \in S$ defined by:

$$h(u, v) = \frac{(v + u)}{\|(v + u)\|},$$

in which $v \in S$ is the output direction, and $u \in S$ is the light incident direction (Figure 9.4). Then we define the Modified Blinn-Phong's model by:

$$f(x, u, v) = k_s \langle n, h(u, v) \rangle^p + k_d,$$

such that $p \in \mathbb{N}$ controls the spread of light made the surface. The light is becomes more concentrated in the reflection direction when we increase $p$.

The explanation for the use of the $h$ in the Blinn-Phong's model can be found in [31].

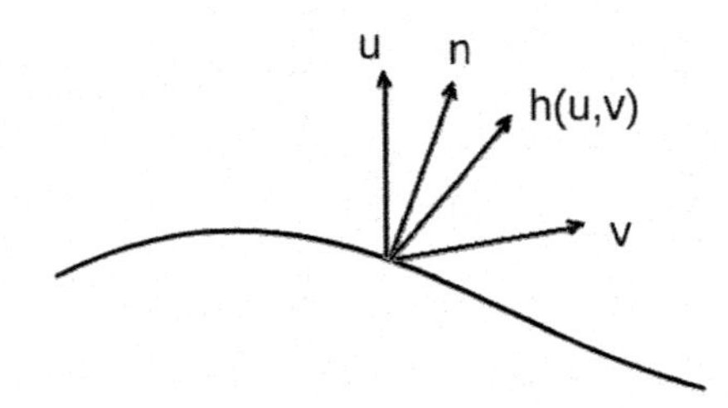


Figure 9.4 Vectors used in the definition of Blinn-Phong's model.

## 9.6 NUMERICAL APPROXIMATION

The Rendering Equation 9.1 can be written in a more compact way as:

$$L = L_e + TL, \tag{9.4}$$

such that $T$ is the functional defined in the function space by:

$$T(g(x,v)) = \int_{S_x} f(x,u,v)\langle u,n\rangle g(x,u)\,du$$

.

One way to get an approximate solution for the Equation 9.4 is to substitute $L$ for the expression on the right side. For example, doing one substitution we get:

$$L = L_e + T(L_e + TL) = L_e + TL_e + T^2 L$$

Repeating the process $n+1$ times we get:

$$L = L_e + \sum_{i=1}^{n} T^n L_e + T^{n+1} L$$

Ignoring the residual term of order $n+1$ this relationship gives a way to get an approximation from the solution of the Rendering Equation 9.4:

$$L = L_e + \sum_{i=1}^{n} T^n L_e$$

If we consider all $n \in \mathbb{N}$ we get a Newman Series [17], and it can be physically understand as the fact that the light emitted from a point in the scene is defined by the light emitted by this point in the direction of the observer plus all the light reflected by that point considering $1, 2, 3, \ldots$ reflections in the scene. This is the base of the Monte Carlo solution for the Rendering Equation.

## 9.7 MONTE CARLO INTEGRATION METHOD

The Monte Carlo is an algorithm that finds the approximation for an integral using a probabilistic process. In order to understand this algorithm we firstly must know the definition of Expected Value $E[x]$ of a probabilistic variable with density function defined by $p(x)$ in the set $\Omega$.

We define $E[x]$ by:

$$E[x] = \int_{x\in\Omega} xp(x)\,dx$$

The idea consists in explore the Law of Large Numbers [16], which states that: When $n \to \infty$:

$$E[x] \approx \frac{1}{n}\sum_{i=1}^{n} x_i,$$

such that $x_1, x_2, \ldots, x_n$ are samples of a probabilistic variable with density $p(x)$.

More precisely, the Law of Large Number states that if all $x_i$ are random variables independent and identically distributed then [27]

$$Probability\left[E(x) = \lim_{n\to\infty}\frac{1}{n}\sum_{i=1}^{n} x_i\right] = 1.$$

We can also define the expected value of a function $f(x)$ by:

$$E[f(x)] = \int_{x\in\Omega} f(x)p(x)\,dx$$

It means that if we want to use the Expected Value integration as an approximation of

$$\int_{x\in\Omega} f(x)\,dx$$

we may define a variable with probabilistic density $g$ given by

$$g(x) = \frac{f(x)}{p(x)},$$

In this case, when $n \to \infty$ :

$$\int_{x\in\Omega} f(x)\,dx = \int_{x\in\Omega} g(x)p(x)\,dx = E(g(x)) \approx \frac{1}{n}\sum_{i=1}^{n} g(x_i),$$

such that $x_1, x_2, \ldots, x_n$ are samples of a probabilistic variable with density $p(x)$.

## 9.8 PATH TRACING

Let us consider a scene with many pure reflective surfaces and a luminary.

The Path Tracing algorithm consists on applying the Monte Carlo Integration Method to solve the Rendering Equation exploring the Numerical Approximation explained in the Section 9.6.

As explained previously, if we truncate the Newman Series we get an approximated solution for the Rendering Equation. Each term of this summation consists on an integral that can be solved using the Monte Carlo Method.

More precisely, in order to get a sample that approximates the Rendering Equation in a pixel $p$, we can trace a ray starting in the camera center and whose direction corresponds to $p$. After that, every time we get a intersection with a surface in the scene we create a new ray with a uniform distribution in the hemisphere over the surface. We repeat this process until a ray intersects the luminarie (Figure 9.5).

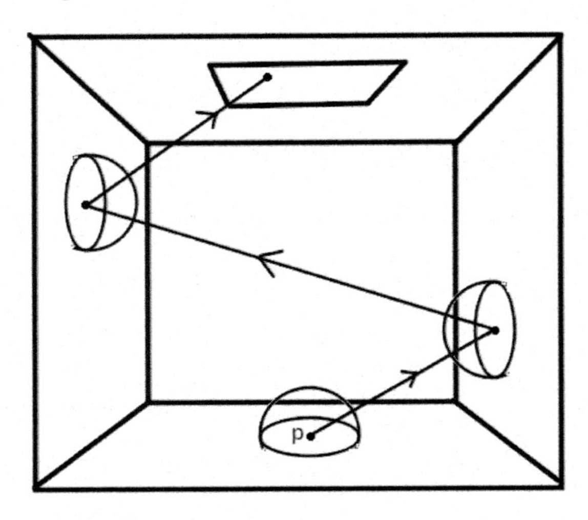


Figure 9.5 Sampling a path relative to the pixel $p$.

We apply the BRDF in each intersection to estimate the amount of light each surface is reflecting to the next surface. Each path give us an approximation to the Newman Series that solves the Rendering Equation. Calculating the mean value obtained from many paths in the same pixel we increase the accuracy of the Rendering Equation solution on it.

## 9.9 UNIFORM SAMPLING OVER A HEMISPHERE

There is an easy way to get a uniform sample of a hemisphere over a surface with normal $n \in \mathbb{R}^3$ using a conditional probability process.

Firstly we choose a point $x_i \in \mathbb{R}^3$ inside the cube $[-1,1] \times [-1,1] \times [-1,1]$ with a uniform probability. After that, we discard $x_i$ if $\|x_i\| > 1$. Finally if $\langle x_i, n \rangle > 0$ we accept $x_i/\|x_i\|$ as a sample, otherwise we accept $-x_i/\|x_i\|$ as a sample (Figure 9.6).

## 9.10 SPLITTING THE DIRECT AND INDIRECT ILLUMINATION

In the algorithm presented on Section 9.8 the paths may be very large until they hits the luminaries. A better way to get an approximation to the solution of the Rendering Equation consists on splitting the direct and the indirect illumination on a 3D scene.

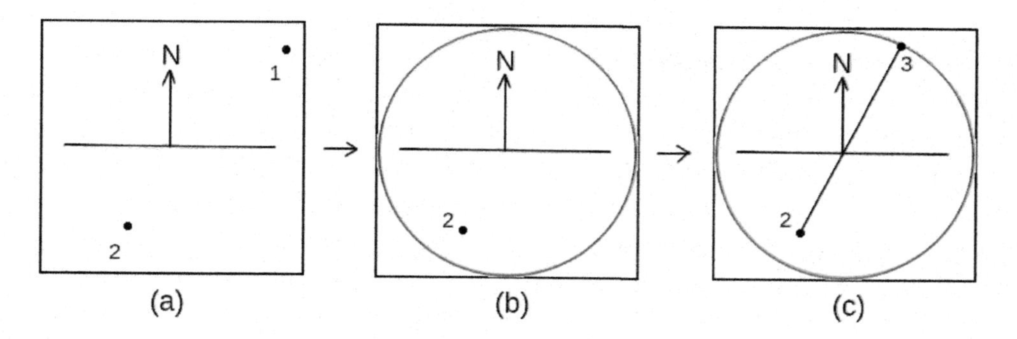


Figure 9.6 (a) Two points uniform sampled inside the cube $[-1,1] \times [-1,1] \times [-1,1]$. (b) Discarding the point outside the unitary sphere. (c) Getting the normalized point such that $\langle x_i, n \rangle > 0$.

In other words:

$$TotalRadiance = DirectRadiance + IndirectRadiance$$

More precisely, we consider on each intersection that the light comes from two different direction, one of them corresponding to a reflection from a ray traced to another surface and the other from a set of rays traced to points chosen in the luminaries (Figure 9.7).

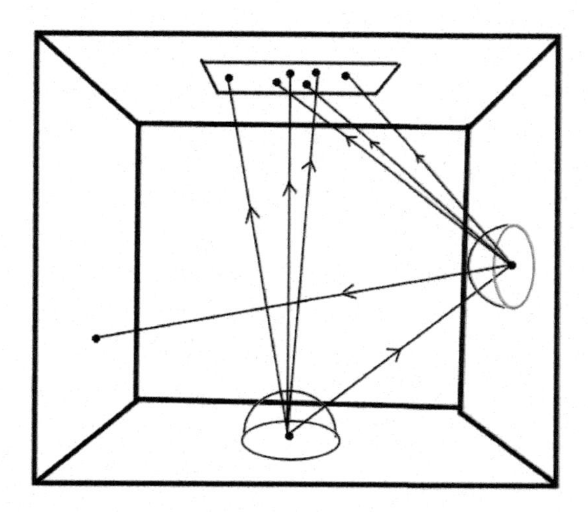

Figure 9.7 Splitting the radiance evaluation into two parts: the direct and the indirect illumination. The indirect illumination is estimated by one uniform random ray chosen on a hemisphere and the direct illumination is estimated by many rays traced to the luminarie.

Of course, the surface of the luminaries must be excluded from the scene if a reflective ray hits it. Thus in this case, it must be considered a zero illumination, since the direct light is being calculated apart. Another important fact is that the surfaces that are perfect mirror must only use the indirect light model. In order to calculate the direct light we may use the radiance expressed in the form of Section 9.4. More precisely, the radiance on a point $x$ of a surface in the scene coming from the luminarie can be estimated by a Monte Carlo algorithm in the following way.

1. We choose a random point $x'$ over the luminarie following a uniform sampling process.
2. Following the definitions of $V$ and $G$ in Section 9.4, we get a term of a Monte Carlo sum given by terms of the form:

$$L(x,u,v) = \frac{f(x,u,v)L_e(x,u)V(x,x')G(x,x')}{p(x')} =$$
$$f(x,u,v)L_e(x,u)V(x,x')G(x,x')A,$$

such that $A$ is the area of the luminary.

As a conclusion we get that an approximation for the direct light coming from a luminary is given by:

$$L(x,v) \approx \frac{1}{n}\sum_{i=1}^{n} f(x,u_i,v)L_e(x,u_i)V(x,x_i')G(x,x_i')A, \tag{9.5}$$

in which $x_1', x_2', \ldots, x_n'$ are points sampled over the surface of the luminary following a uniform distribution, $u_i \in \mathbb{R}^3$ is defined by:

$$u_i = \frac{x_i' - x}{\|x_i' - x\|},$$

and the functions $G$ and $V$ are the ones defined in the Section 9.4.

## 9.11 POLYGONAL LUMINARIES

In order to define the direct illumination generated from a polygonal light using the Equation 9.5 we need to know the area of the polygonal and a process to sample this polygon in a uniform way.

Since luminaries will not be used in visual effects we will just consider the case in which they can be splitted into triangles with the same area. In this case, we can just uniformly sample each triangle.

If $v_0$, $v_1$, $v_2$ are vertices of a triangle, and $r_1$,$r_2$ are two uniform random variables on the interval $[0,1]$. Then we can obtain a uniform random point $(x,y,z)$ over the triangle if we choose $x$, $y$ and $z$ of the form [20]:

$$x = (1-\sqrt{r_1})v_0$$
$$y = \sqrt{r_1}(1-r_2)v_1$$
$$z = r_2\sqrt{r_1}v_2$$

We can calculate the polygon area by adding the area of those triangles each one calculated by the following expression [12]:

$$\frac{1}{2}\|(v_2 - v_0) \times (v_1 - v_0)\|.$$

## 9.12 CODE MODULES

We extended the code given in the book [12] implementing a Path Tracing code. Some modules are included in the "base" directory of the library of this book.

### 9.12.1 Path Tracing API

These functions are part of the folder /base/lptrace. Another set of functions of the same folder will be analyzed on the next chapter.

```
Color trace_path( int first_level, int level, Ray v,
                  Object *ol );
```

Recursive code that implements the Path Tracing algorithm. The *first_level* receives the number of inter-reflections between the scene surfaces. The *level* represents the current reflection deep it reduces one unit at each recursion. $v$ represent the current ray which we want to estimate the color. *ol* is the list of all objects in the scene. In this function if the *se* parameter of the material of some object is negative then the path tracing acts like a mirror reflection.

```
Color mirror_reflect( int first_level, int level, Ray v,
         Inode *i, Vector3 p, Object *ol, Material *m );
```

This function returns the color in the perfect mirror direction. The *first_level*, *level*, $v$, and *ol* have the same interpretation of the *trace_path* function. $i$ receives the inode relative to the point $p$, that correspond to the point where we are calculating the reflection over the surface. $m$ corresponds to the material in the point $p$.

```
Color apply_bphong( Material *m, Color lcolor, Vector3 n,
                    Vector3 l, Vector3 v, Real geomfactor );
```

This function implements the BRDF based on the Blinn-Phong reflection model described in the Section 9.5.3. $m$ receives the material description using the same structure defined in the S3D library. *lcolor* defines the color of the incident light. $n$ receives the normal direction of the surface. $l$ receives the incident light direction. $v$ receives the view direction. *geomfactor* is used to allows this function to be used in the case of polygonal lights or in the case of light bouncing over regular surfaces. In the case of regular surfaces

$$geomfactor = 2\pi\langle n, l\rangle.$$

On the other hand, if the surface is a polygonal light *geomfactor* is defined by the value of Equation 9.3 multiplied by the area of the polygon.

```
Vector3 sort_new_direction( Cone c );
```

This function returns a uniform sampling over the hemisphere described by $c$, implementing the algorithm of the Section 9.9.

```
Bool hit_surface( Ray v, Object *ol );
```

This function returns TRUE if the ray $v$ crosses a object in the object list $ol$, and FALSE otherwise.

```
void adjust_normal( Inode *i, Ray v );
```

This function makes the normal $i \to n$ of the inode $i$ corresponds to a vector that points to the half-space that contains the ray $v$.

### 9.12.2 Path Tracing Code

```
// extracted from ibl/ibl.h

#define REAL_RANDOM   ((double)rand()/RAND_MAX)

// extracted from lptrace/ptrace.h

#ifndef PTRACE_H
#define PTRACE_H

#include <math.h>
#include "rshade.h"
#include "rt.h"
#include "ibl.h"
#include "plight.h"
#include <time.h>
#include <stdlib.h>
#include <string.h>


Color trace_path( int first_level, int level, Ray v, Object *ol );
Color mirror_reflect( int first_level, int level, Ray v, Inode *i,
                      Vector3 p, Object *ol, Material *m );
Color apply_bphong( Material *m, Color lcolor, Vector3 n, Vector3 l,  Vector3 v,
                    Real geom_fac );
Vector3 sort_new_direction( Cone c );
Bool hit_surface( Ray v, Object *ol );
void adjust_normal( Inode *i, Ray v );

#endif


// lptrace/ptrace.c

#include "ptrace.h"

static Color direct_light( int ns, Vector3 p, Vector3 n, Vector3 v,
                           Object *ol, PolyLight *l, Material *m );
```

```
static int light_visible(Ray r, Object *ol, PolyLight *l, Color *c );

#include "ptrace.h"


Color trace_path( int first_level, int level, Ray v, Object *ol )
{
 PolyLight *l = plight_head;
 Color c = C_BLACK, c_aux, w;
 Ray r;
 Inode *i = ray_intersect(ol, v);

 adjust_normal( i, v );
 if( light_visible(v,ol,l,&w) )
   return w;

 if((i != NULL) && level){
    level--;
    Material *m = i->m;
    Vector3 p = ray_point(v, i->t);
    Cone  recv = cone_make(p, i->n, PIOVER2);

    if( m->se < 0 )
       c = mirror_reflect( first_level, level, v, i, p, ol, m );
    else{
       Vector3 d = sort_new_direction(recv);
       c = direct_light( DIRECT_LIGHT_NSMPLS, p, i->n,
                         v3_scale(-1,v.d), ol, l, m);
       if( !light_visible(ray_make(recv.o, d), ol, l, &w) ){
         c_aux = trace_path( first_level, level, ray_make(p, d), ol );
         c = c_add(c, apply_bphong( m, c_aux, i->n,
                   d, v3_scale(-1,v.d), 2*PI*v3_dot(d, i->n) ) );
       }
    }
    inode_free(i);
    return c;
 }
 else{
   if( i == NULL )
      return C_BLACK;
   else{
      inode_free(i);
      return C_BLACK;
   }
 }
}

Color direct_light( int ns, Vector3 p, Vector3 n, Vector3 v,
                    Object *ol, PolyLight *l, Material *m )
{
 int k;
 Vector3 r, light_ptdir, inv_light_ptdir;
 Ray s;
 Color c, cl_light;
 Inode *i;
 double g=0;

 c = c_make(0,0,0);
 for( k = 0; k < ns; k++ ){
   s = plight_sample(l,&cl_light);
   r = v3_sub(s.o, p);
   light_ptdir = v3_unit(r);
   inv_light_ptdir = v3_scale( -1, light_ptdir );
   i = ray_intersect(ol, ray_make(p, light_ptdir));
```

```
   if( (i == NULL) || (i->t > v3_norm(r)) ){
      g = fabs( v3_dot(inv_light_ptdir,s.d)*v3_dot(light_ptdir, n) );
      c = c_add(c, apply_bphong(m, cl_light, n, light_ptdir, v,
                  g*plight_area(l)/SQR(v3_norm(r) ) ) );
   }
   if( i != NULL )
      inode_free(i);
 }

 return v3_scale(1./ns, c);
}

int light_visible(Ray r, Object *ol, PolyLight *l, Color *c)
{
 Inode *i1, *i2;

 i1 = plight_intersect( l, r, c );
 if( i1 != NULL ){
   i2 = ray_intersect( ol, r );
   if( i2 == NULL || (i1->t < i2->t ) ){
     inode_free(i1);
     if( i2 != NULL )
       inode_free(i2);
     return TRUE;
   }
   inode_free(i1);
   if( i2 != NULL )
     inode_free(i2);
 }
 return FALSE;
}


// Extracted from lptrace/aux.c

#include "ptrace.h"


Color mirror_reflect( int first_level, int level, Ray v, Inode *i,
                      Vector3 p, Object *ol, Material *m )
{
 Vector3 d, c_aux;

 d = v3_sub( v.d, v3_scale( 2*v3_dot( v.d, i->n ), i->n ) );
 c_aux = trace_path( first_level, level, ray_make(p, d), ol );
 return v3_scale( m->ks, c_aux );
}

Color apply_bphong( Material *m, Color lcolor, Vector3 n, Vector3 l,
                    Vector3 v, Real geom_fac)
{
 Color cs,cd;
 Vector3 h = v3_unit(v3_add(l,v));

 cs = c_scale(m->ks* pow(MAX(0, v3_dot(h,n)),m->se)*geom_fac, lcolor);
 cd = c_mult(m->c, c_scale(m->kd*geom_fac, lcolor) );
 return c_add( cs, cd );
}


Vector3 sort_new_direction( Cone c )
{
 double x, y, z;
 Vector3 v, n = v3_unit(c.d);
```

```
  do{
     x = 2*REAL_RANDOM - 1;
     y = 2*REAL_RANDOM - 1;
     z = 2*REAL_RANDOM - 1;
     v = v3_make(x,y,z);
  }
  while( v3_norm(v) > 1 );

  if( v3_dot(v, c.d) < 0 ){
      double s = v3_dot(v3_scale(-1,v), n);
      v = v3_add( v, v3_scale(2*s,n) );
  }
  return v3_unit(v);
}


Bool hit_surface( Ray v, Object *ol )
{
  Inode *i = ray_intersect(ol, v);
  if( i == NULL )
     return FALSE;
  else{
     inode_free(i);
     return TRUE;
  }
}


void adjust_normal( Inode *i, Ray v )
{
  if( i != NULL && v3_dot( i->n, v.d ) > 0 )
      i->n = v3_scale( -1, i->n );
}
```

### 9.12.3 Poly Light API

```
void init_plight_list( void );
```

This function initializes an empty list of polygonal lights.

```
PolyLight *plight_alloc( Color c, Hpoly *p );
```

This function returns a polygonal light with color $c$ and shape defined by $p$.

```
void plight_insert( PolyLight *pl );
```

This function inserts the *pl* in the polygonal lights list.

```
void plight_free( PolyLight *head );
```

This function erases the list of polygonal lights pointed by *head*.

```
Ray plight_sample( PolyLight *p, Color *c );
```

This function returns a ray whose origin is a uniform sample of the polygonal light $p$, and whose direction is normal to $p$. $c$ receives the color of the sample.

The direction of the returned ray is important because it is used to evaluate the light transport described in the $G$ function described by 9.3.

```
Real plight_area( PolyLight *p );
```

This function returns the area of the polygonal light $p$.

```
Inode *plight_intersect( PolyLight *p, Ray r, Color *c );
```

This function returns the inode that correspondent to the intersection of the ray $r$ with the polygonal light $p$. $c$ returns the color of the light, and if there is not an intersection the function returns $NULL$.

```
Val plight_parse(int pass, Pval *pl);
```

This function is responsible to process the polygonal light in the scene description file.

```
Ray poly3_sample( Hpoly *p );
```

This function returns a ray whose origin is a uniform sample of the polygonal light $p$, and whose direction is normal to $p$.

### 9.12.4 Poly Light Code

```
// plight/plight.h

#ifndef P_LIGHT_H
#define P_LIGHT_H

#define DIRECT_LIGHT_NSMPLS 3

#include "color.h"
#include "poly.h"
#include "ibl.h"


typedef struct PolyLight{
  struct PolyLight *next;
  Hpoly *p;
  Color c;
} PolyLight;
```

```
PolyLight *plight_head;

void init_plight_list( void );
void plight_free( PolyLight *head );
PolyLight *plight_alloc( Color c, Hpoly *p );
void plight_insert( PolyLight *pl );
Ray plight_sample( PolyLight *p, Color *c );
Real plight_area( PolyLight *p );
Inode *plight_intersect( PolyLight *p, Ray r, Color *c );
Val plight_parse(int pass, Pval *pl);

Ray poly3_sample( Hpoly *p );

#endif
```

```
// plight/plight.c

#include "plight.h"


void init_plight_list( void )
{
 plight_head = NULL;
}


PolyLight *plight_alloc( Color c, Hpoly *p )
{
 PolyLight *pl;

 pl = NEWSTRUCT(PolyLight);
 pl->next = NULL;
 pl->p = p;
 pl->c = c;
 return pl;
}


void plight_free( PolyLight *head )
{
 PolyLight *pl;

 for( pl = head; pl != NULL; pl=pl->next ){
   free( pl->p );
   free( pl );
 }
}


void plight_insert( PolyLight *pl )
{
 pl->next = plight_head;
 plight_head = pl;
}


Ray plight_sample( PolyLight *pl, Color *c )
{
 int i, r, n = 0;
 PolyLight *aux;

 for( aux = pl; aux != NULL; n++, aux = aux->next );
 r = rand()%n;
```

```
 for( aux = pl, i=0; i < r-1; aux = aux->next );

 *c = aux->c;
 return poly3_sample(aux->p);
}


Real plight_area( PolyLight *pl )
{
 Real a = 0;

 while (pl != NULL) {
  a += hpoly3_area(pl->p);
  pl = pl->next;
 }
 return a;
}


Inode *plight_intersect( PolyLight *pl, Ray r, Color *c )
{
 Inode *l = NULL;

 while( pl != NULL && l == NULL ){
   l = hpoly_intersect(pl->p, hpoly3_plane(pl->p), r);
   *c = pl->c;
   pl = pl->next;
 }
 return l;
}


Val plight_parse(int pass, Pval *plist)
{
 Color c;
 struct Hpoly *pl, *pols;
 Val v = {V_NULL, 0};

 if (pass == T_EXEC) {
   Pval *p;

   for (p = plist; p !=NULL; p = p->next) {
     if (p->name == NULL) {
       error("(extense light) syntax error");
     } else if (strcmp(p->name, "color") == 0) {
         c = pvl_to_v3(p->val.u.v);
     } else if (strcmp(p->name, "shape") == 0 && p->val.type == V_HPOLYLIST) {
         pols = p->val.u.v;
     } else error("(extense light) syntax error");
   }
   for(pl = pols; pl != NULL; pl = pl->next)
      plight_insert( plight_alloc( c, pl ));
   v.type = V_PL_LIGHT;
   v.u.v = NULL;
  }
  return v;
}


Ray poly3_sample( Hpoly *p )
{
 Vector3 a, b, c, v0, v1, v2;
 Real r1 = REAL_RANDOM;
 Real r2 = REAL_RANDOM;
```

```
  v0 = v3_make(p->v[0].x, p->v[0].y, p->v[0].z);
  v1 = v3_make(p->v[1].x, p->v[1].y, p->v[1].z);
  v2 = v3_make(p->v[2].x, p->v[2].y, p->v[2].z);

  a = v3_scale(1 - sqrt(r1), v0);
  b = v3_scale(sqrt(r1)*(1 - r2), v1);
  c = v3_scale(sqrt(r1)*r2, v2);

  return ray_make( v3_add(a, v3_add(b, c)), hpoly_normal(p));
}
```

## 9.13 RENDERING PROGRAM

This is the code of a basic path tracing algorithm, prepared for rendering a synthetic image from a scene description using the extension of the language defined in the S3D library. This extension allows the definition of polygonal luminaries.

In the code, the macros $MAX\_PTRACE\_SAMPLES$ and $PATH\_SIZE$ define, respectively, the number of paths traced from each pixel and the deep of the path used.

```
// ptrace/main.h

#include <stdio.h>
#include <stdlib.h>
#include <math.h>
#include <ctype.h>

#include "image.h"
#include "defs.h"
#include "geom.h"
#include "stack.h"
#include "view.h"
#include "poly.h"
#include "prim.h"
#include "hier.h"
#include "lang.h"
#include "clip.h"
#include "raster.h"
#include "shade.h"
#include "ray.h"
#include "csg.h"
#include "rt.h"
#include "rshade.h"

#include "scene.h"
#include "ptrace.h"
#include "ibl.h"

void init_render(Scene *s);
Ray ray_view(int u, int v);
void init_lang(void);

// ptrace/main.c

#include "main.h"

static Scene *s;
static Matrix4 mclip, mdpy;
```

```
#define MAX_PTRACE_SAMPLES 200
#define PATH_SIZE 3

int main(int argc, char **argv)
{
  Color c, c_aux;
  int u, v, smpl, ll_y, ur_y;
  Ray r;

  init_lang();
  s = scene_read();
  init_render(s);
  s->objs = graph_flatten(graph_transform(s->objs));

  ll_y = s->view->sc.ll.y;
  ur_y = s->view->sc.ur.y;

#pragma omp parallel for private(u, r, c, smpl, c_aux) shared(s)
           schedule(dynamic, 1)
  for (v = ll_y; v < ur_y; v += 1) {
    for (u = s->view->sc.ll.x; u < s->view->sc.ur.x; u += 1) {
      r = ray_unit(ray_transform(ray_view(u, v), mclip));
      c = c_make(0,0,0);
      for( smpl = 0; smpl < MAX_PTRACE_SAMPLES; smpl++ ){
         c_aux = trace_path(PATH_SIZE, PATH_SIZE, r, s->objs, c_make(0,0,0) );
         c = c_add( c, c_aux );
      }
      img_putc(s->img, u, v,
               col_dpymap( c_scale(1./MAX_PTRACE_SAMPLES, c))
      );
    }
  }
  img_write(s->img,"stdout",0);
  exit(0);
}

Ray ray_view(int u, int v)
{
  Vector4 w = v4_m4mult(v4_make(u, v,s->view->sc.ur.z, 1), mdpy);
  return ray_make(v3_v4conv(v4_m4mult(v4_make(0, 0, 1, 0), mdpy)),
  v3_make(w.x, w.y, w.z));
}

void init_render(Scene *s)
{
  mclip = m4_m4prod(s->view->Vinv, s->view->Cinv);
  mdpy = m4_m4prod(s->view->Pinv, s->view->Sinv);
}

void init_lang(void)
{
  lang_defun("scene", scene_parse);
  lang_defun("view", view_parse);
  lang_defun("dist_light", distlight_parse);
  lang_defun("plastic", plastic_parse);
  lang_defun("primobj", obj_parse);
  lang_defun("sphere", sphere_parse);
  lang_defun("polyobj", obj_parse);
  lang_defun("trilist", htrilist_parse);
  lang_defun("group", group_parse);
  lang_defun("hdrdome", hdrdome_parse);
  lang_defun("polylight", plight_parse);
}
```

## 9.14 RESULT

As an example, considering the following scene description, the main program generates as output the image of Figure 9.8.

Figure 9.8 Image generated using the Path Tracing software, defining the macro *MAX PTRACE SAMPLES* as 2000.

```
scene{
       polylight{  color = {10,10,10},
                   shape = trilist {
                   {{3.3,3.3,9.9999}, {6.6,3.3,9.9999}, {3.3,6.6,9.9999}},
                   {{3.3,6.6,9.9999}, {6.6,3.3,9.9999}, {6.6,6.6,9.9999}}
                }
       },

       camera = view {
          from = {5, 20, 5}, at = {5, 0, 5}, up = {0,0,1}, fov = 60
          imgw = 640, imgh = 480 },

       object = primobj{
       material = plastic { kd = 0.4, ks = 0.9, kt =0, se = -1
                  d_col = {1, 1, 1}, s_col = {1,1,1}},
                  shape = sphere { center = {2, 3, 2}, radius = 2}},

       object = primobj{
       material = plastic { kd = 0.4, ks = 0.3, kt = 0,
                  d_col = {.9, .9, .6}, s_col = {1,1,1}},
                  shape = sphere { center = {6, 6, 2}, radius = 2}},

       object = primobj{
       material = plastic { kd = 0.4, ks = 0.3, kt = 0,
                  d_col = {1, 1, 1}, s_col = {1,1,1}},
                  shape = sphere { center = {3, 6, .9}, radius = .9}},

       object = polyobj {
       material = plastic { kd = .4, ks = 0, kt = 0,
                  d_col = {.9, .9, .9}, s_col = {1,1,1}},
                  shape = trilist {
```

```
                    {{0, 0, 0},  {10, 0, 0}, {0, 10, 0}},
                    {{10, 10, 0}, {0, 10, 0}, {10, 0, 0}}
    }},

    object = polyobj {
    material = plastic { kd = .4, ks = 0, kt = 0,
               d_col = {.9, .9, .9}, s_col = {1,1,1}},
               shape = trilist {
                    {{0, 0, 10},  {0, 10, 10}, {10, 0, 10}},
                    {{10, 10, 10}, {10, 0, 10}, {0, 10, 10}}
    }},

    object = polyobj {
    material = plastic { kd = .4, ks = 0, kt = 0,
               d_col = {.6, .6, .9}, s_col = {1,1,1}},
               shape = trilist {
                    {{0,0,0},  {0,10,0}, {0,10, 10}},
                  {{0,0,0}, {0,10,10}, {0,0,10}}
    }},

    object = polyobj {
    material = plastic { kd = .4, ks = 0, kt = 0,
               d_col = {.9, .6, .6}, s_col = {1,1,1}},
               shape = trilist {
                    {{10,0,0},  {10,10,10}, {10,10, 0}},
                    {{10,0,0}, {10,0,10}, {10,10,10}}
    }},

    object = polyobj {
    material = plastic { kd = .4, ks = 0, kt = 0,
               d_col = {.9, .9, .9}, s_col = {1,1,1}},
               shape = trilist {
                    {{0,0,0},  {10,0,10}, {10,0, 0}},
                    {{0,0,0}, {0,0,10}, {10,0,10}}
    }},
};
```

CHAPTER 10

# Image-Based Lighting

NOW WE WILL PRESENT how to create a 3D Scene that has the illumination compatibilized with the real world illumination. In order to do this we will need to capture a radiance map, which defines the amount or radiance coming from each direction. This radiance map is then sampled when we render the scene.

Firstly we will describe an algorithm that is able to acquire a picture whose pixel's values are proportional to the radiance. After that, we will explain how this picture can be used to generate a spherical map of radiance. In the sequence, we will explain how to generate an image rendered with global illumination whose illumination is compatible to the real scene. Finally we will explain how estimate the BRDF of a surface designed to capture the shadows.

## 10.1 HDR PICTURES

Whenever we capture a picture, the values of each pixel usually do not correspond to a value proportional to the incoming radiance on that pixel. Besides that, depending on the exposure time of the camera, the image can have regions with saturated color values or to dark color values.

In order to adjust the illumination of a virtual scene to the illumination of a real scene, we need to know the amount of radiance coming from each direction. This kind of information can be stored in an image whose pixels have a High-Dynamic Range (HDR). Which means that the pixels of the images can stores a very high contrast stored as a floating point data.

The next sections will explain how to combine a set of regular pictures acquired with different exposure times to reconstruct the relationship between the pixel values and radiance values.

## 10.2 RECONSTRUCTING THE HDR RADIANCE MAP

Although the captured value in each pixel is not proportional to the radiance we can use the reciprocity equation in order to estimate it, such as is explained in [2].

This equation establish a relationship between the irradiance $E_i$ received by the pixel sensor and pixels values by $Z_{ij}$, in which $i$ denotes a spatial index over pixels

DOI: 10.1201/9781003206026-10

and $j$ denotes different exposure times $\Delta t_j$. More precisely this equation is expressed by:

$$Z_{ij} = f(E_i \Delta t_j). \tag{10.1}$$

Assuming that $f$ is monotonic it can be invertible, and we can rewrite this equation as

$$f^{-1}(Z_{ij}) = E_i \Delta t_j.$$

Applying the logarithm on both sides of this equation, we have:

$$lnf^{-1}(Z_{ij}) = lnE_i + ln\Delta t_j.$$

Defining $g = lnf^{-1}$ we conclude that:

$$g(Z_{ij}) = lnE_i + ln\Delta t_j. \tag{10.2}$$

We have that $\Delta t_j$ can be defined in the camera and $Z_{ij}$ can be measured in the image. As a consequence, the only unknown values are $E_i$ and all the values assumed by the function $g$.

We can recover the function $g$ and the irradiance $E_i$ by solving the Equation 10.2 by a optimization process. More precisely, we can do this minimizing the objective function:

$$\sum_{i=1}^{N} \sum_{j=1}^{P} (g(Z_{ij}) - lnE_i + ln\Delta t_j)^2 + \lambda \sum_{z=Z_{min}+1}^{Z_{max}-1} g''(z)^2, \tag{10.3}$$

such that $N$ is the number of pixels locations, $P$ is the number of pictures, $Z_{min}$ is the smallest pixel value and $Z_{max}$ is the greatest pixel value. In this equation, the first summation intends to establish the relationship defined by the Equation 10.2 and the second one is a term used to ensure that $g$ is smooth. Besides that, the $g''(z)$ may be approximated to this discrete version $g(z-1) - 2g(z) + g(z+1)$. The $\lambda$ controls the amount smoothness of the estimated function $g$.

This objective function defines a linear least square problem that can be solved using the algorithm presented in the Section 3.2.

It is necessary to add an additional constraint to this problem, since $g(z)$ and $E_i$ are defined up to a scale factor $\alpha$. In other words, if $lnE_i$ is replaced by $lnE_i + \alpha$, and $g$ is replaced by $g + \alpha$ then the value of the objective function is unchanged.

In order to solve this problem, we can add a constraint to the optimization problem. More precisely, we can use the constraint $g(Z_{mid}) = 0$, such that $Z_{mid} = \frac{1}{2}[Z_{min} + Z_{max}]$.

Another problem that must be considered is the fact that the function $g$ can have a steep slope near $Z_{min}$ and $Z_{max}$. As a consequence, we can expect that it is can be less smooth neat those values. In order to explore this fact in the objective function, we can define a weighting function $w(z)$ to emphasize the values of the middle of the $g$ function. More precisely we can define $w$ as:

$$w(z) = z - Z_{min},\ for\ z \leqslant Z_{mid},$$

and

$$w(z) = Z_{max} - z,\ for\ z > Z_{mid}.$$

The objective function can be improved by the formulation:

$$\sum_{i=1}^{N}\sum_{j=1}^{P}[w(Z_{ij})(g(Z_{ij}) - lnE_i + ln\Delta t_j)]^2 + \lambda \sum_{z=Z_{min}+1}^{Z_{max}-1} [w(z)g(z-1) - 2g(z) + g(z+1)]^2 + g(Z_{mid})^2 \tag{10.4}$$

## 10.3 COLORED PICTURES

For solving the problem in the case of colored pictures we can apply the process explained before in order to find the radiance related to each pixel sensor. Unfortunately, there exist a scale ambiguity relating those values. Using the process explained before, we will have that the RGB value $(Z_{mid}, Z_{mid}, Z_{mid})$ will have equal radiance value for R, G and B, which means that the pixel is achromatic.

## 10.4 RECOVERING AN HDR RADIANCE MAP

After solving the optimization problem of the previous section we recover the $g$ function, and we are able to use it to recover the incident radiance $E_i$ on each sensor of the camera by any exposure time $\Delta t_j$. More precisely, we have that

$$lnE_i = \frac{\sum_{j=1}^{P} w(Z_{ij})(g(Z_{ij}) - ln\Delta t_j)}{\sum_{j=1}^{P} w(Z_{ij}).}$$

## 10.5 THE PFM FILE FORMAT

The HDR pictures can be stored by PFM (Portable FloatMap) files. This type was designed by Paul Debevec following the spirit of the Portable Pixmap format (.ppm).

The header of a PFM file is formed by strings with the following data:

```
[type]
[xres][yres]
[byte order]
```

The lines of the header end with a 0x0a Unix character, which means a carriage return.

If the file stores a RGB HDR picture the [*type*] is the string "*PF*", and when it is monochrome it is "*Pf*".

$[xres] \times [yres]$ is the resolution of the image.

Finally the [*byte order*] stores a string $''1.0''$ if the pixels values are stored in a big-endian style and $''-1.0''$ if they are stored in little-endian order.

After the header, there is a sequence of IEEE 754 single precision floating points storing each channel of the pixels sequence represented specified in left to right, bottom to top order.

## 10.6 CONVERSIONS BETWEEN LDR AND HDR IMAGES

Our Image-Based Lighting software processes only HDR images. In order to use it, is necessary to convert the background video stored in low-dynamic range (LDR) images into a set of PFM images, generating the input of our software. We also need to convert the PFM images generated as the output into a regular video composed by LDR images. To make these conversions, we can use the command *convert* available in the Image Magick software.

If the name of the input video is video.mp4, we can generate the input frames using the command:

```
convert video.mp4 -endian lsb bk%d.pfm
```

To create the output video with $n$ frames we can use the command of the form:

```
convert out%d.pfm[0-n] output.mp4
```

## 10.7 FROM HDR PICTURES TO EQUIRECTANGULAR PROJECTIONS

In order to capture the radiance of a scene, we can take several pictures of a mirror ball using different exposure times. It can be done by a setup like Figure 10.1. For

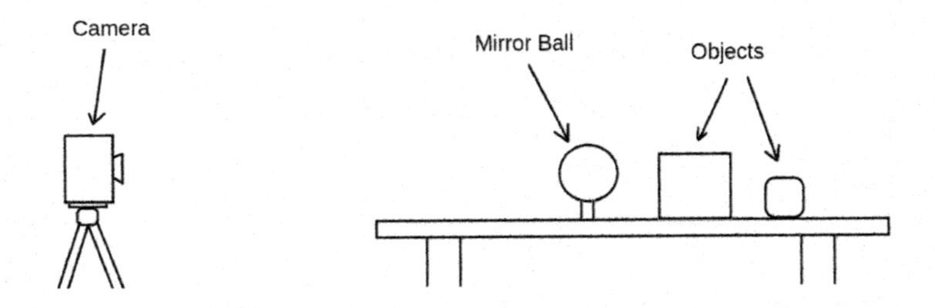


Figure 10.1 Setup used for capturing a radiance map.

example, we captured a set of pictures of a mirror ball such as presented in Figure 10.2.

By doing this, each pixel of projected camera correspond to a sampling of a direction in the 3D scene, as illustrated in the Figure 10.3.

Figure 10.2 Pictures captured from a mirror ball with different exposure times.

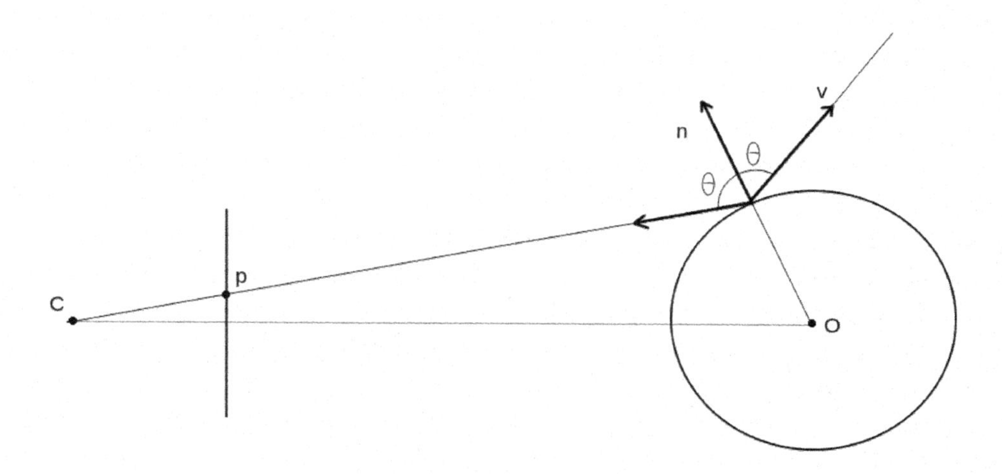


Figure 10.3 The projection of a mirror ball captured by a camera whose optical center is $C$. The color of the direction $v$ is encoded in the pixel $p$.

If we assume that the camera is far away from the mirror ball and zooming in, we can consider that the camera works approximately as a orthogonal projection camera, and the rays captured by them are almost parallel. In this situation only a small part of the scene is not captured in the picture taken by the camera (Figure 10.4).

Each ray can be mapped into a value of latitude and longitude. In our case, we assume that the latitude is measured as the angle defined to the south pole of the sphere (Figure 10.5), and the longitude is measured inside the interval $[-\pi, \pi]$.

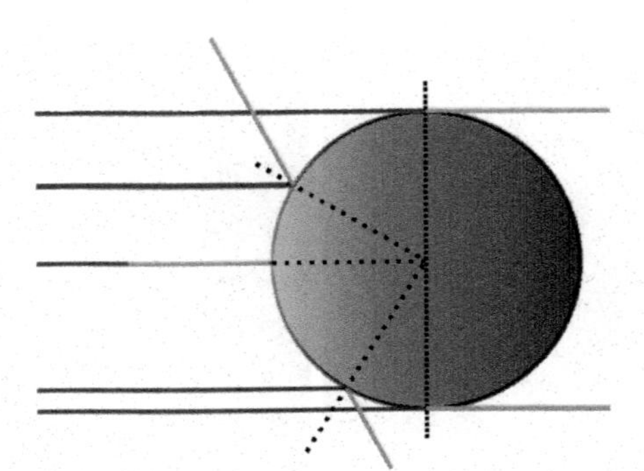


Figure 10.4 How the rays of the world are reflected into mirror ball assuming an orthographic camera. The blue rays are the reflections of the green scene rays.

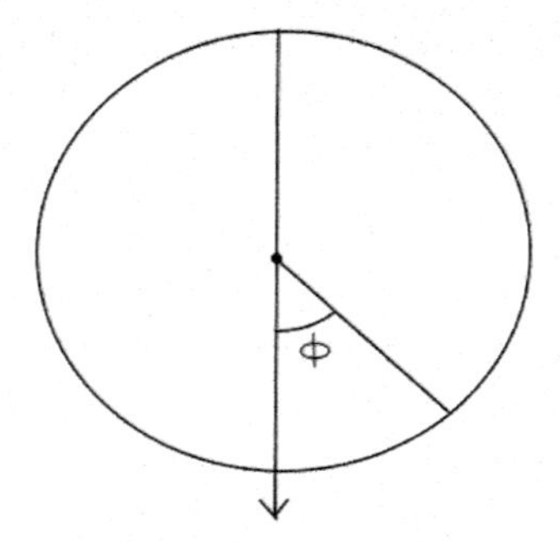


Figure 10.5 Measuring the latitude relative to the south pole.

It means that the rays are mapped to a rectangle $[-\pi, \pi] \times [0, \pi]$. After that, this rectangle is uniformly sampled on a grid, generating an image with resolution $2w \times w$, such that $w \in \mathbb{N}$. This kind of mapping is known as equirectangular projection.

We did not implement a software for processing pictures of mirror balls. Instead, we used the HDR Luminance software for generating the radiance equirectangular map and used the program Gimp to adjust the exposure (Figure 10.6).

## 10.8 ORIENTING THE RADIANCE DOME

In our solution we decided to deal with the radiance equirectangular map as a far away dome that works as a source of light of the scene, considering it as a dome whose radius is infinity. This is an approximation, since this radiance comes from surfaces in the scene. On the other hand, it is easier to be assumed and usually produces high-quality results.

We also decided not to consider the radiance as a separated direct light. Instead, we decided to use its radiance as the source of light of a path that ends tracing a ray into a direction that does not intersect any surface of the scene (Figure 10.7).

Figure 10.6 Equirectangular radiance map generated using the softwares HDR Luminance and Gimp.

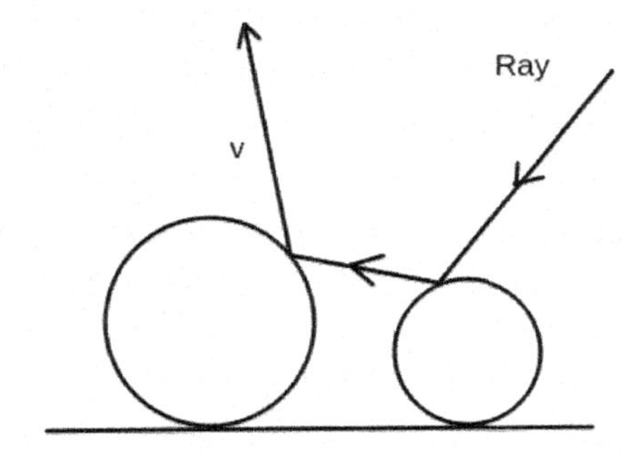


Figure 10.7 An example of a path that after two reflections it does not intersect any surface in the scene. The vector $v$ must be used to determine the sample in the radiance dome.

In order to map rays into pixels in the equirectangular radiance map it is necessary to define the dome orientation accordingly to the MatchMove referential system. To do this, we need to find the extrinsic parameters of the camera used for capturing the mirror ball.

A possible solution for this problem uses the parameters of a camera calibrated by the correspondence of a 3D point cloud, which is estimated in the MatchMove coordinate system, and the respective 2D projections made by the camera that captured the mirror ball without zoom. The 3D point cloud can be estimated by using the programs presented in the Chapter 8, and the camera calibration can be performed by the programs presented in the Sections 4.11 or 4.12. The second program has the advantage of demanding less correspondences to be used, and our experience shows that it is precise enough for the purpose of orienting the radiance dome.

As a result, we find a camera $k[R|t]$. The origin $o \in \mathbb{R}^3$ of the longitude coordinate system over the mirror ball is then given by the opposite of the view direction, which means that:

$$o = -R^T(0,0,1)^T,$$

and the origin of the latitude coordinate system $s \in \mathbb{R}^3$ is given by the direction related to the y-axis of the captured image. In other words:

$$s = R^T(0,1,0)^T$$

We must define another vector in order to define a coordinate system. Since $o$ and $s$ are unitary and orthogonal we can define an unitary vector $w \in \mathbb{R}^3$ orthogonal to both by

$$w = s \times o.$$

## 10.9 RENDERING USING A RADIANCE MAP

After choosing the unitary vector $v$ that does not intersect any object in the scene and that belongs to the ray in the end of a path in the pathtracing algorithm, we must discover the correspondent pixel in the equirectangular radiance map. We can do it using the following process:

1. Orienting the radiance dome as explained in Section 10.8.
2. Define the matrix $M$ that converts coordinates in the scene into coordinates in the dome. The rows of $M$ are $o$, $w$ and $s$, defined in the Section 10.8.

$$M = \begin{pmatrix} o_x & o_y & o_z \\ w_x & w_y & w_z \\ s_x & s_y & s_z \end{pmatrix}.$$

3. Calculate $p = Mv$.
4. Calculate the latitude as $\theta = arccos(p_z)$ and the longitude as $\phi = atan2(p_y, p_x)$, such that $atan2$ has the meaning defined in the $C$ language library.
5. Find the pixel coordinate $(x, y)$ in the equirectangular image by assuming that $x = round((1 - \frac{\phi}{\pi})\frac{w}{2})$ and $y = round(\frac{\theta}{\pi}h)$, in which the HDR image that stores the radiance values has the resolution $w \times h$, and $round$ is a function that returns the closest integer value to the parameter.

Obviously, the matrix $M$ can be calculated only one time, and the steps 1 and 2 can be executed once.

## 10.10 INTERACTION BETWEEN THE REAL AND VIRTUAL SCENES

The essence of Image-Based Lighting has been introduced in [3]. In the approach presented in this article, the scene was split into three parts: the Distant Scene, the Local Scene and the Synthetic Objects (Figure 10.8-a). The Distant Scene consists of objects far away from the synthetic objects. This part provides light to the Local Scene and to the Synthetic Objects but not receives light from them. This contribution corresponds to the radiance map recovered using the mirror ball. The Local Scene correspond to the real objects that are close to the synthetic ones and that have a large interaction with them. In our code this objects correspond to a polygon that can be used to catch the shadow of the synthetic objects over a support plane. This polygon must have their real BRDF modeled in order to be used by the pathtracing software. The Synthetic Objects are the ones created by the user. They have their geometry and BRDF defined by the user.

In our model, we decided to use a little different approach, in which the synthetic object does not receive light from the local scene (Figure 10.8-b). We decided to do this because the radiance coming from the radiance dome has an approximation of the reflected radiance from the local scene, and this is a simpler solution.

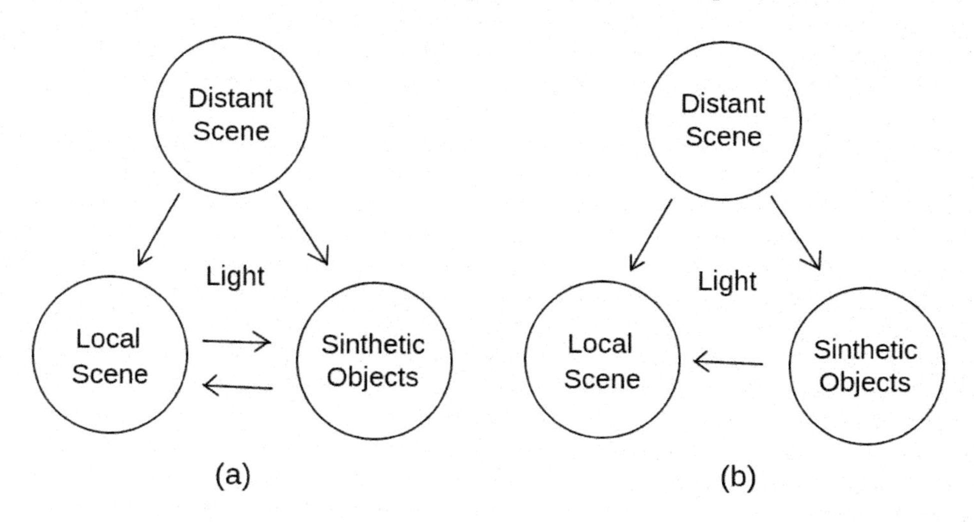


Figure 10.8 (a) Image-Based Lighting approach introduced by Paul Debevec in [3]. (b) Image-Based Lighting approach in our render software.

### 10.10.1 Modeling the BRDF of the Local Scene

In our software we considered that the local surface is a polygon that has a perfect Lambertian BRDF with a constant color. Although this approximation seems to be very restrictive it is not the case, as will become clear in the Section 10.10.2.

We can estimate the BRDF firstly rendering the local scene assuming a S3D material with $kd = 1$ and $d\_col = (1, 1, 1)$. Then we can adjust the color comparing the output image rendered image with these parameters with the color of the background image used as the input. More precisely, if $(R, G, B)$ is the sum of the colors of all the pixels of the local scene background image, and $(R', G', B')$ is the sum of the colors of all pixels of the local scene rendered image, then we can model the *d_col* of the local scene polygon as

$$d\_col = \left(\frac{R}{R'}, \frac{G}{G'}, \frac{B}{B'}\right),$$

and the $k_d$ coefficient must keep set with the value 1.

### 10.10.2 Differential Rendering

The differential rendering is a procedure introduced in [3] for rendering the local scene. It generates acceptable results even when the local scene in modeled with a constant BRDF although they have a diffuse reflection with some textured appearance in the background image.

This procedure consists on assuming that the color of the points in the local scene is calculated subtracting from the local scene background image the difference between the rendered local scene without the presence of synthetic objects and the rendered of the local scene in the presence of synthetic objects. Synthetically it can be expressed by:

$$LS_{final} = LS_b - (LS_{nobj} - LS_{obj})$$

Or, in other words:

$$LS_{final} = LS_b + (LS_{obj} - LS_{nobj}),$$

such that $LS_{final}$ is the color of local scene final image, $LS_b$ is the color of local scene in the background image, $LS_{obj}$ is the color of the local scene rendered interacting with the synthetic objects, and $LS_{nobj}$ is the local scene rendered without the presence of synthetic objects.

## 10.11 CODE MODULES

### 10.11.1 HDR Image API

```
HDRImage *hdrimg_init(int w, int h);
```

This function creates an HDR image with resolution $w \times h$.

```
void hdrimg_free( HDRImage *img );
```

This function deallocates the image *img*.

```
void hdrimg_write(HDRImage *i, char *fname);
```

This function save the HDR image $i$ into a PFM file whose name is *fname*.

```
HDRImage *hdrimg_read(char *fname);
```

This function returns a HDR image represented by the PFM file whose name is *fname*.

```
Color hdrimg_getc( HDRImage *p, int u, int v );
```

This function returns the colored radiance encoded in the pixel $(u, v)$ assuming coordinates defined in relation to a bottom left origin.

```
void hdrimg_putc( HDRImage *p, int u, int v, Color c );
```

This function changes the color radiance $c$ in the pixel with coordinates $(u, v)$ in the HDR image $p$. The coordinates considers an origin in the left bottom corner of the image.

### 10.11.2 HDR Image Code

```
// ibl/hdr.h

#ifndef HDR_H
#define HDR_H

#include "color.h"
#include "image.h"
#include "string.h"


typedef struct HDRImage{
  int w, h;
  Color *c;
}HDRImage;


HDRImage *hdrimg_init( int w, int h);
void hdrimg_free( HDRImage *img );
void hdrimg_write(HDRImage *i, char *fname);
HDRImage *hdrimg_read(char *fname);

Color hdrimg_getc( HDRImage *p, int u, int v );
void hdrimg_putc( HDRImage *p, int u, int v, Color c );

#endif
```

```
// ibl/hdr.c

#include "hdr.h"


static void read_line( FILE *f, char *s );


HDRImage *hdrimg_init( int w, int h )
{
 HDRImage *img = NEWSTRUCT(HDRImage);

 img->w = w;
```

```
 img->h = h;
 img->c = NEWTARRAY( w*h, Color );
 return img;
}


void hdrimg_free( HDRImage *img )
{
 free( img->c );
 free( img );
}


Color hdrimg_getc( HDRImage *img, int x, int y )
{
 return img->c[img->w*y + x];
}


void hdrimg_putc( HDRImage *img, int x, int y, Color c )
{
 img->c[img->w*y + x] = c;
}


void hdrimg_write(HDRImage *img, char *fname)
{
 Color c;
 float cx, cy, cz;
 char s[10];
 int x, y, xres, yres;
 FILE *f = fopen( fname, "wb" );

 fwrite( "PF\n", 3, 1, f );
 sprintf( s, "%i %i\n", img->w, img->h );
 fwrite( s, strlen(s), 1, f );
 sprintf( s, "-1.0\n" );
 fwrite( s, strlen(s), 1, f );
 for ( y = 0; y < img->h; y++ )
   for ( x = 0; x < img->w; x++ ){
     c = hdrimg_getc( img, x, y );
     cx = (float)c.x;
     fwrite( &cx, sizeof(float), 1, f );
     cy = (float)c.y;
     fwrite( &cy, sizeof(float), 1, f );
     cz = (float)c.z;
     fwrite( &cz, sizeof(float), 1, f );
 HDRImage *img;
 int x, y, xres, yres;
   }
 fclose(f);
}


HDRImage *hdrimg_read(char *fname)
{
 float r, g, b, v;
 char s[100];
 FILE *f = fopen( fname, "rb" );

 read_line( f, s );
 if( strcmp( s, "PF\n" ) )
   error("wrong PFM File Format\n" );
 read_line( f, s );
```

```
  sscanf( s, "%d %d\n", &xres, &yres );
  read_line( f, s );
  sscanf( s, "%f", &v );
  if( v != -1 )
    error("wrong PFM File Format\n" );
  img = hdrimg_init( xres, yres );

  if( v == -1 ){
    for( y = 0; y < img->h; y++ )
      for( x = 0; x < img->w; x++ ){
        fread( &r, sizeof(float), 1, f );
        fread( &g, sizeof(float), 1, f );
        fread( &b, sizeof(float), 1, f );
        hdrimg_putc( img, x, y, c_make(r, g, b) );
      }
  }
  fclose(f);
  return img;
}


void read_line( FILE *f, char *s )
{
  int i = 0;
  do{
    fread( &s[i], 1, 1, f );
  }
  while( s[i++] != 0x0A );
  s[i] = 0x00;
}
```

### 10.11.3 Image-Based Light API

```
void hdrdome_init( void );
```

This function initializes the HDR dome of the scene with the NULL value.

```
HDRDome *hdrdome_alloc( Vector3 orig, Vector3 south,
                        HDRImage *img );
```

This function creates a radiance dome whose orientation is specified by the parameters. This orientation relates the dome coordinates with the correspondent equirectangular radiance map as explained in the Section 10.9. The vector *south* specifies the axis of the sphere pointing to the floor defining the latitude coordinate. The vector *orig* specifies the origin used for defining the longitude coordinate. Finally, the HDR image *img* defines the equirectangular radiance map used by the dome.

```
void hdrdome_free( HDRDome *d );
```

This function deallocates the radiance dome specified by *d*.

```
Color hdrdome_value( HDRDome *d, Vector3 v );
```

This function gets the trichromatic HDR color of the dome $d$ in the direction $v$.

```
Val hdrdome_parse(int pass, Pval *pl);
```

This function implements the parsing function used by the scene description language used by HDR domes.

```
PolyShadow *plshadow_alloc( Material *m, Hpoly *p,
                            HDRImage *img );
```

This function creates a PolyShadow with geometry specified by $p$, material specified by $m$ and background image specified by $img$.

```
void plshadow_free( PolyShadow *ps );
```

This function destroys the PolyShadow ps.

```
Val plshadow_parse(int pass, Pval *plist);
```

This function is responsible for parsing a PolyShadow in the scene description language.

```
Object *plshadow_to_obj( PolyShadow *ps );
```

This function converts the PolyShadow ps into an object.

### 10.11.4 Image-Based Light Code

```
// ibl/ibl.h

#ifndef IBL_H
#define IBL_H


#include "color.h"
#include "hdr.h"
#include "lang.h"
#include "shade.h"
#include "poly.h"
#include "obj.h"
#include <string.h>

#define REAL_RANDOM    ((double)rand()/RAND_MAX)
```

```
typedef struct HDRDome{
  Vector3 u;
  Vector3 v;
  Vector3 n;
  HDRImage *img;
}HDRDome;

typedef struct PolyShadow{
  struct PolyShadow *next;
  Hpoly *p;
  Material *m;
  HDRImage *img;
}PolyShadow;

HDRDome *hdr_dome;

/* HDR Dome Functions */

void hdrdome_init( void );
HDRDome *hdrdome_alloc( Vector3 orig, Vector3 south,
                        HDRImage *img );
void hdrdome_free( HDRDome *d );
Color hdrdome_value( HDRDome *d, Vector3 v );
Val hdrdome_parse(int pass, Pval *pl);


/* Poly Shadow Functions */

PolyShadow *plshadow_alloc( Material *m, Hpoly *p, HDRImage *img );
void plshadow_free( PolyShadow *ps );
Val plshadow_parse(int pass, Pval *plist);
Object *plshadow_to_obj( PolyShadow *ps );


#endif

// ibl/ibl.c

#include "ibl.h"


static Color hdrmap_value( HDRDome *d, Vector3 v );
static Vector3 change_ref( Vector3 v, HDRDome *d );


void hdrdome_init( void )
{
 hdr_dome = NULL;
}


HDRDome *hdrdome_alloc( Vector3 orig, Vector3 south, HDRImage *img )
{
 HDRDome *d = NEWSTRUCT( HDRDome );

 d->n = v3_unit( south );
 d->u = v3_unit( v3_sub( orig,
                 v3_scale( v3_dot( orig, d->n ), d->n )) );
 d->v = v3_cross( d->n, d->u );
 d->img = img;

 return d;
}
```

```
void hdrdome_free( HDRDome *d )
{
 hdrimg_free( d->img );
 free( d );
}


Color hdrdome_value( HDRDome *d, Vector3 v )
{
 Real theta, phi;

 v = change_ref( v, d );
 theta = acos(v.z);
 phi = atan2(v.y,v.x);

 return hdrimg_getc( d->img, ROUND((1 - (phi/PI))*d->img->w/2 ),
                             ROUND((theta/PI)*d->img->h) );
}


Vector3 change_ref( Vector3 v, HDRDome *d )
{
 Matrix4 m;

 m = m4_ident();
 m.r1.x = d->u.x;  m.r1.y = d->v.x;  m.r1.z = d->n.x;
 m.r2.x = d->u.y;  m.r2.y = d->v.y;  m.r2.z = d->n.y;
 m.r3.x = d->u.z;  m.r3.y = d->v.z;  m.r3.z = d->n.z;

 return v3_m3mult( v, m );
}

// ibl/plshadow.c

#include "ibl.h"


PolyShadow *plshadow_alloc( Material *m, Hpoly *p, HDRImage *img )
{
 PolyShadow *ps;

 ps = NEWSTRUCT(PolyShadow);
 ps->m = m;
 ps->p = p;
 ps->img = img;
 return ps;
}


void plshadow_free( PolyShadow *ps )
{
 free( ps->m );
 free( ps->p );
 free( ps->img );
 free( ps );
}


Object *plshadow_to_obj( PolyShadow *ps )
{
 Shape *sh = shape_new(V_HPOLYLIST, ps->p);
 Object *o = obj_new(V_HPOLYLIST, sh);
```

```
 o->mat = ps->m;

 return o;
}

// Extracted from lang/sdltypes.h

#define V_PL_LIGHT 501
#define V_HDR_DOME 502
#define V_HDR_SCENE 503
#define V_PL_SHADOW 504


// ibl/lang.c

#include "ibl.h"
#include "sdltypes.h"


Val hdrdome_parse(int pass, Pval *pl)
{
  Pval *p;
  Val v = {V_NULL, 0};

  if (pass == T_EXEC) {
    Vector3 o, s;
    HDRDome *dome;

    o = pvl_get_v3(pl, "orig", v3_make(1,0,0));
    s = pvl_get_v3(pl, "south", v3_make(0,-1,0));
    p = pl;
    while (p != NULL) {
      if (p->name && strcmp(p->name,"hdrimg") == 0 && p->val.type == V_STR)
dome = hdrdome_alloc( o, s, hdrimg_read( (char*)p->val.u.v ) );
      p = p->next;
    }

    v.type = V_HDR_DOME;
    v.u.v = dome;
    hdr_dome = (HDRDome*)dome;
  }
  return v;
}


Val plshadow_parse(int pass, Pval *plist)
{
 struct Hpoly *pols;
 Material *m;
 HDRImage *img = NULL;
 PolyShadow *plshadow;
 Val v = {V_NULL, 0};

 if (pass == T_EXEC) {
   Pval *p;

   for (p = plist; p !=NULL; p = p->next) {
     if (p->name == NULL) {
       error("(shadow) syntax error");
     } else if (strcmp(p->name, "material") == 0 && p->val.type == V_MATERIAL) {
         m  = p->val.u.v;
     } else if (strcmp(p->name, "shape") == 0 && p->val.type == V_HPOLYLIST) {
         pols = p->val.u.v;
     } else if( strcmp(p->name,"img") == 0 && p->val.type == V_STR ){
```

```
        img = hdrimg_read( (char*)p->val.u.v );
      } else error("(shadow) syntax error");
    }
    plshadow = plshadow_alloc( m, pols, img );
    v.type = V_PL_SHADOW;
    v.u.v = plshadow;
  }
  return v;
}
```

### 10.11.5 HDR Scene API

```
Val hdrscene_parse(int c, Pval *pl);
HDRScene *hdrscene_read(void);
HDRScene *hdrscene_eval(void);
void hdrscene_free(HDRScene *s);
```

These functions are very similar to the correspondent scene functions in the S3D Library. The unique difference is that they are ready to parse an HDRDome and a PolyShadow structures, and the image structure generated after parsing the scene has the type HDRImage.

### 10.11.6 HDR Scene Code

```
// hdrscene/hdrscene.h

#ifndef HDR_SCENE_H
#define HDR_SCENE_H

#include <stdio.h>
#include "lang.h"
#include "obj.h"
#include "sdltypes.h"
#include "view.h"
#include "scene.h"
#include "hdr.h"
#include "ibl.h"
#include "plight.h"
#include "ibl.h"

typedef struct HDRScene {
  struct View   *view;
  struct HDRImage  *hdrimg;
  struct HDRDome   *hdrdome;
  struct PolyLight *plights;
  struct PolyShadow *plshadow;
  struct Object *objs;
} HDRScene;


Val hdrscene_parse(int c, Pval *pl);
HDRScene *hdrscene_read(void);
HDRScene *hdrscene_eval(void);
void hdrscene_free(HDRScene *s);

#endif
```

```
// hdrscene/hdrscene.c

#include "hdrscene.h"


static struct View *view = NULL;
static struct HDRImage *hdrimg = NULL;
static struct HDRDome  *hdrdome = NULL;
static struct PolyLight *plights = NULL;
static struct PolyShadow *plshadow = NULL;
static struct Object *objs = NULL;

#ifndef LIST_INSERT
#define LIST_INSERT(L, I, TYPE) {struct TYPE *tt = I; tt->next = L; L = tt; }
#endif

static void hdrcollect_items(Pval *pl)
{
  Pval *p = pl;
  while (p != NULL) {
    if (p->name && strcmp(p->name,"object") == 0 && p->val.type == V_OBJECT)
      objs = obj_insert(objs, p->val.u.v);
    else if (p->name && strcmp(p->name,"object") == 0 && p->val.type == V_GROUP)
      objs = obj_list_insert(objs, p->val.u.v);
    else if (p->name && strcmp(p->name,"camera") == 0 && p->val.type == V_CAMERA)
      view = p->val.u.v;
    else if (p->name && strcmp(p->name,"dome") == 0 && p->val.type == V_HDR_DOME)
      hdrdome = p->val.u.v;
    else if (p->name && strcmp(p->name,"light") == 0 && p->val.type == V_PL_LIGHT)
      LIST_INSERT(plights, p->val.u.v, PolyLight)
    else if(p->name && strcmp(p->name,"shadow") == 0 && p->val.type == V_PL_SHADOW)
      LIST_INSERT(plshadow, p->val.u.v, PolyShadow)
    p = p->next;
  }
}


Val hdrscene_parse(int c, Pval *pl)
{
  Val v = {V_NULL, 0};

  if (c == T_EXEC) {
    HDRScene *s = NEWSTRUCT(HDRScene);
    view = NULL; hdrimg = NULL; plights = NULL; objs = NULL;

    hdrcollect_items(pl);
    s->view = view;
    s->hdrimg = hdrimg;
    s->hdrdome = hdrdome;
    s->objs = objs;
    s->plights = plights;
    s->plshadow = plshadow;
    v.type = V_HDR_SCENE;
    v.u.v = s;
  }
  return v;
}


HDRScene *hdrscene_read(void)
{
  if (lang_parse() == 0)
    return hdrscene_eval();
  else
```

```
    error("(hdrscene read)");
}


HDRScene *hdrscene_eval(void)
{
  HDRScene *s;
  Val v = lang_nd_eval();
  if (v.type != V_HDR_SCENE)
    error("(hdrscene eval)");
  else
    s =  v.u.v;
  if (s->view == NULL)
    s->view = initview();
  if (s->hdrimg == NULL)
    s->hdrimg = hdrimg_init( s->view->sc.ur.x, s->view->sc.ur.y );
  return s;
}

void hdrscene_free(HDRScene *s)
{
  if (s->objs)
    obj_list_free(s->objs);
  if (s->view)
    efree(s->view);
  if (s->hdrimg)
    hdrimg_free(s->hdrimg);
  if (s->hdrdome)
    hdrdome_free(s->hdrdome);
  if (s->plights)
    plight_free(s->plights);
  if(s->plshadow)
    plshadow_free(s->plshadow);
  efree(s);
}
```

### 10.11.7 Dome Path Tracing API

```
Color dome_trace_path( int first_level, int level, Ray v,
                       Object *ol );
```

Recursive code that implements the Path Tracing algorithm with a radiance dome surrounding the scene. The *first_level* receives the number of inter-reflections between the scene surfaces. The *level* represents the current reflection deep. It reduces one unit at each recursion. $v$ represents the current ray which we want to estimate the color. *ol* is the list of all objects in the scene. In this function if the *se* parameter of the material of some object is negative then the path tracing acts like a mirror reflection.

```
Color dome_mirror_reflect( int first_level, int level, Ray v,
                Inode *i, Vector3 p, Object *ol,
                Material *m );
```

This function returns the color in the perfect mirror direction. The $first\_level$, $level$, $v$, and $ol$ have the same interpretation in the $dome\_trace\_path$ function. $i$ receives the inode relative to the point $p$, that correspond to the point where we are calculating the reflection over the surface. $m$ corresponds to the material in the point $p$.

### 10.11.8 Dome Path Tracing Code

```
// Extracted from lptrace/ptrace.h


Color dome_trace_path( int first_level, int level, Ray v, Object *ol );
Color dome_mirror_reflect( int first_level, int level, Ray v, Inode *i,
                           Vector3 p, Object *ol, Material *m );

// lptrace/dmptrace.c

#include "ptrace.h"


Color dome_trace_path( int first_level, int level, Ray v, Object *ol )
{
 HDRDome *dome = hdr_dome;
 Color c = C_BLACK, c_aux, w;
 Ray r;
 Inode *i = ray_intersect(ol, v);

 adjust_normal( i, v );
 if((i != NULL) && level){
    level--;
    Material *m = i->m;
    Vector3 p = ray_point(v, i->t);
    Cone  recv = cone_make(p, i->n, PIOVER2);
    Vector3 d = sort_new_direction(recv);

    if( m->se < 0 )
       c = dome_mirror_reflect( first_level, level, v, i, p, ol, m );
    else{
       c_aux = dome_trace_path( first_level, level, ray_make(p, d), ol );
       c = apply_bphong( m, c_aux, i->n,  d, v3_scale(-1,v.d),
                         2*PI*v3_dot(d, i->n) );
    }
    inode_free(i);
    return c;
 }
 else{
   if(( i == NULL ) && (level == first_level ) ){
      return C_BLACK;
   }
   else{
      if( i == NULL )
        return hdrdome_value( dome, v.d );
      if( i != NULL )
        inode_free(i);
      return C_BLACK;
   }
 }
}


// Extracted from lptrace/aux.c
```

```
Color dome_mirror_reflect( int first_level, int level, Ray v, Inode *i,
                           Vector3 p, Object *ol, Material *m )
{
 Vector3 d, c_aux;

 d = v3_sub( v.d, v3_scale( 2*v3_dot( v.d, i->n ), i->n ) );
 c_aux = dome_trace_path( first_level, level, ray_make(p, d), ol );
 return v3_scale( m->ks, c_aux );
}
```

## 10.12 POLYSHADOW COLOR ADJUST PROGRAM

This program can be used to adjust the parameter *d_color* of a polygon that catches shadows in the scene, such as explained in Section 10.10.1.

In order to correctly use this program the correspondent "plshadow"command in the scene description must be set with $kd = 1$ and any value to *d_col*. After running the program it will print on the screen the value that must be use as a substitution of the *d_col* vector. We must keep *kd* with the value 1.

$argv[1]$ receives the name of the file containing the intrinsic parameters of the camera.

$argv[2]$ receives the name of the file that contains the extrinsic parameters related to each frame.

$argv[3]$ receives the number of frame that must be used to run the program. It is important to choose a frame in that there is not any object occluding the polygon that catches the shadow.

*stdin* receives the scene encoded in the scene description language.

The program assumes that the video that must be processed have been split into frames whose names are *bk%d.pfm* in which *%d* is an integer. It can be done as explained in Section 10.6.

```
// getcolor/main.h

#include <stdio.h>
#include <stdlib.h>
#include <math.h>
#include <ctype.h>

#include "image.h"
#include "defs.h"
#include "geom.h"
#include "stack.h"
#include "view.h"
#include "poly.h"
#include "prim.h"
#include "hier.h"
#include "lang.h"
#include "clip.h"
#include "raster.h"
#include "shade.h"
#include "ray.h"
#include "csg.h"
#include "rt.h"
#include "rshade.h"

#include "hdrscene.h"
```

```
#include "ptrace.h"
#include "ibl.h"
#include "mmove.h"

void init_render(void);
Ray ray_view(int u, int v);
void init_lang(void);

// getcolor/main.c

#include "main.h"

static HDRScene *s;
static Matrix4 mclip, mdpy;

#define MAX_NFRAMES 1000
#define MAX_PTRACE_SAMPLES  100
#define PATH_SIZE 1
#define MAX_N 10000

void main( int argc, char **argv )
{
  Real sl;
  Color bk, c_aux, c_med, bk_med;
  int u, v, smpl, ll_y, ur_y;
  long int n;
  Ray r;
  Object *obj;
  CamData *cd;
  HDRImage *aux;
  char str[50];

  srand(time(NULL));
  hdrdome_init();
  init_lang();
  s = hdrscene_read();
  aux = hdrimg_read( "bk0.pfm" );
  cd = cam_data_alloc( MAX_NFRAMES, aux->w, aux->h, .01, 1000. );
  mmove_read( cd, argv[1], argv[2] );
  mmove_view( s->view, cd, atoi( argv[3] ) );
  init_render();

  sprintf( str, "bk%i.pfm", atoi( argv[3] ) );
  s->plshadow->img = hdrimg_read( str );

  ll_y = s->view->sc.ll.y;
  ur_y = s->view->sc.ur.y;

  obj = plshadow_to_obj( s->plshadow );

  n = 0;
  c_med = C_BLACK;
  bk_med = C_BLACK;
  for(v = ll_y; v < ur_y; v += 1) {
     for (u = s->view->sc.ll.x; u < s->view->sc.ur.x; u += 1) {
       r = ray_unit(ray_transform(ray_view(u, v), mclip));
       bk = hdrimg_getc( s->plshadow->img, u, v );
       for( smpl = 0; smpl < MAX_PTRACE_SAMPLES; smpl++ ){
          if( (hit_surface( r, obj )) && (!hit_surface( r, s->objs )) &&
              (s->plshadow != NULL) && (s->plshadow->img != NULL) &&
              (n < MAX_N) ){
                 c_aux = dome_trace_path(PATH_SIZE, PATH_SIZE, r, obj );
                 c_med = v3_add( v3_scale( ((Real)n)/(n+1), c_med ),
                                 v3_scale( 1./(n+1), c_aux ));
```

```
                    bk_med = v3_add( v3_scale( ((Real)n)/(n+1), bk_med ),
                                    v3_scale( 1./(n+1), bk ));
                    n++;
                }
            }
        }
    }

 if(s->plshadow != NULL)
   printf( "%lf %lf %lf\n", (bk_med.x/c_med.x)*s->plshadow->m->c.x,
                            (bk_med.y/c_med.y)*s->plshadow->m->c.y,
                            (bk_med.z/c_med.z)*s->plshadow->m->c.z );
}


Ray ray_view(int u, int v)
{
  Vector4 w = v4_m4mult(v4_make(u, v,s->view->sc.ur.z, 1), mdpy);
  return ray_make(v3_v4conv(v4_m4mult(v4_make(0, 0, 1, 0), mdpy)),
  v3_make(w.x, w.y, w.z));
}

void init_render(void)
{
  mclip = m4_m4prod(s->view->Vinv, s->view->Cinv);
  mdpy = m4_m4prod(s->view->Pinv, s->view->Sinv);
}

void init_lang(void)
{
  lang_defun("hdrscene", hdrscene_parse);
  lang_defun("hdrdome", hdrdome_parse);
  lang_defun("view", view_parse);
  lang_defun("plastic", plastic_parse);
  lang_defun("primobj", obj_parse);
  lang_defun("sphere", sphere_parse);
  lang_defun("polyobj", obj_parse);
  lang_defun("trilist", htrilist_parse);
  lang_defun("group", group_parse);
  lang_defun("plshadow", plshadow_parse );
}
```

## 10.13 VISUAL EFFECTS PROGRAM

Now we present the program that combines the image-based lighting, pathtracing and matchmove to generate a visual effect. This program can be used to generate all the frames of the video or a single image.

$argv[1]$ receives the name of the file containing the intrinsic parameters of the camera.

$argv[2]$ receives the name of the file that contains the extrinsic parametes related to each frame.

$argv[3]$ is an optional parameter that receives the number of the frame that must be rendered. If we omit this parameter the program renders the entire video.

*stdin* receives the scene encoded in the scene description language.

In the code, the macros $MAX\_PTRACE\_SAMPLES$ and $PATH\_SIZE$ define, respectively, the number of paths traced from each pixel and the deep of the path used.

The program assumes that the video that must be processed has been split into frames whose names are $bk\%d.pfm$ in which $\%d$ is an integer. The output frames is made in pictures of the form $out\%d.pfm$ such that $\%d$ is the frame number.

We can convert a video in the input and the output into a video as explained in Section 10.6.

```
// mmove/main.h

#include <stdio.h>
#include <stdlib.h>
#include <math.h>
#include <ctype.h>

#include "image.h"
#include "defs.h"
#include "geom.h"
#include "stack.h"
#include "view.h"
#include "poly.h"
#include "prim.h"
#include "hier.h"
#include "lang.h"
#include "clip.h"
#include "raster.h"
#include "shade.h"
#include "ray.h"
#include "csg.h"
#include "rt.h"
#include "rshade.h"

#include "hdrscene.h"
#include "ptrace.h"
#include "ibl.h"
#include "mmove.h"

void render_frame( HDRScene *s );
Ray ray_view(int u, int v);
void init_render(void);
void init_lang(void);
int get_ncameras( FILE *f );

// mmove/main.c

#include "main.h"


static HDRScene *s;
static Matrix4 mclip, mdpy;

#define MAX_NFRAMES 1000
#define MAX_PTRACE_SAMPLES 1000
#define PATH_SIZE 1

Color pfmimg_getc( HDRImage *img, int u, int v  );
void pfmimg_putc( HDRImage *img, int u, int v, Color c );
```

```
void main( int argc, char **argv )
{
 int i, first_frame, max_frame;
 char str[100];
 CamData *cd;
 HDRImage *aux;

 hdrdome_init();
 init_lang();
 s = hdrscene_read();
 aux = hdrimg_read( "bk0.pfm" );
 cd = cam_data_alloc( MAX_NFRAMES, aux->w, aux->h, .01, 1000. );
 mmove_read( cd, argv[1], argv[2] );


 if( argc > 3 ){
    first_frame = atoi( argv[3] );
    max_frame = first_frame + 1;
 }
 else{
    first_frame = 0;
    max_frame = cd->nframes;
 }

 for( i = first_frame; i < max_frame; i++ ){
    printf( "frame %i\n", i );
    sprintf( str, "bk%i.pfm", i );
    s->plshadow->img = hdrimg_read( str );
    mmove_view( s->view, cd, i );
    render_frame( s );
    sprintf( str, "out%d.pfm", i );
    hdrimg_write( s->hdrimg, str );
    hdrimg_free( s->plshadow->img );
 }

 hdrimg_free( aux );
 cam_data_free( cd );
}


void render_frame( HDRScene *s )
{
 Color c, bk, c_aux, c_aux1, c_aux2;
 int u, v, smpl, ll_y, ur_y;
 Ray r;
 Object *obj1, *obj2;

 init_render();
 s->objs = graph_flatten(graph_transform(s->objs));

 ll_y = s->view->sc.ll.y;
 ur_y = s->view->sc.ur.y;

 obj1 = plshadow_to_obj( s->plshadow );
 obj1->next = s->objs;
 obj2 = plshadow_to_obj( s->plshadow );

 #pragma omp parallel for private(u, r, c, smpl, c_aux, c_aux2, c_aux1, bk )\
                         shared(s, obj1, obj2) schedule(dynamic, 1)
  for (v = ll_y; v < ur_y; v += 1) {
    for (u = s->view->sc.ll.x; u < s->view->sc.ur.x; u += 1) {
      r = ray_unit(ray_transform(ray_view(u, v), mclip));
      c = c_make(0,0,0);
      for( smpl = 0; smpl < MAX_PTRACE_SAMPLES; smpl++ ){
```

```
          bk = pfmimg_getc( s->plshadow->img, u, v );
          c_aux = C_BLACK;
          if( (s->plshadow != NULL) && (s->plshadow->img != NULL) )
              c_aux = bk;
          if( hit_surface( r, s->objs ) ){
              c_aux = dome_trace_path(PATH_SIZE, PATH_SIZE, r, s->objs );
          }
          if( (hit_surface( r, obj1 )) && (!hit_surface( r, s->objs )) &&
               (s->plshadow != NULL) && (s->plshadow->img != NULL) ){
                c_aux1 = dome_trace_path(PATH_SIZE, PATH_SIZE, r, obj1 );
                c_aux2 = dome_trace_path(PATH_SIZE, PATH_SIZE, r, obj2 );
                c_aux = v3_add( bk, v3_sub( c_aux1, c_aux2 ) );
          }
          c = c_add( c, c_aux );
      }
      pfmimg_putc( s->hdrimg, u, v, c_scale(1./MAX_PTRACE_SAMPLES, c) );
    }
  }
}


Ray ray_view(int u, int v)
{
  Vector4 w = v4_m4mult(v4_make(u, v,s->view->sc.ur.z, 1), mdpy);
  return ray_make(v3_v4conv(v4_m4mult(v4_make(0, 0, 1, 0), mdpy)),
  v3_make(w.x, w.y, w.z));
}


void init_render(void)
{
  mclip = m4_m4prod(s->view->Vinv, s->view->Cinv);
  mdpy = m4_m4prod(s->view->Pinv, s->view->Sinv);
}


Color pfmimg_getc( HDRImage *img, int u, int v  )
{
 return hdrimg_getc( img, u, img->h - v - 1 );
}


void pfmimg_putc( HDRImage *img, int u, int v, Color c )
{
 hdrimg_putc( img, u, img->h - v - 1, c );
}


void init_lang(void)
{
  lang_defun("hdrscene", hdrscene_parse);
  lang_defun("hdrdome", hdrdome_parse);
  lang_defun("view", view_parse);
  lang_defun("plastic", plastic_parse);
  lang_defun("primobj", obj_parse);
  lang_defun("sphere", sphere_parse);
  lang_defun("polyobj", obj_parse);
  lang_defun("trilist", htrilist_parse);
  lang_defun("group", group_parse);
  lang_defun("plshadow", plshadow_parse );
}
```

## 10.14 RESULTS

There are three images (Figures 10.9, 10.10 and 10.11) from a video rendered with the softwares presented in this chapter. The spheres in the scenes have been inserted into the background images.

Figure 10.9 Frame 2 from a video rendered with the programs described in this chapter.

Figure 10.10 Frame 150 from a video rendered with the programs described in this chapter.

Figure 10.11 Frame 230 from a video rendered with the programs described in this chapter.

The correspondent scene that produced these frames is presented below:

```
hdrscene{

  hdrdome { orig = { 0.415256, 0.459581, -0.785078 },
            south = { 0.093834, 0.836761, 0.539468 },
            hdrimg = "dome.pfm" },

  camera = view { imgw = 1280, imgh = 720 },

  shadow = plshadow{
        material = plastic { kd = 1.0., ks = 0, kt = 0,
          d_col = {0.589600, 0.552094, 0.485632} },
          shape = trilist { {{-3.916377, 1.008441, 10.488646},
                              {-1.723486, 2.929278, 7.749600},
                              {2.102041, -1.828091, 13.835847}},
                             {{2.102041, -1.828091, 13.835847},
                              {-1.723486, 2.929278, 7.749600},
                              {4.294932, 0.092746, 11.096801}}  }
        }

  object = primobj{
         material = plastic { kd = 0, ks = 1.0, kt = 0, se = -1
         d_col = {1, 1, 1}, s_col = {1,1,1}},
         shape = sphere{
           center = {-1.470996, 0.413483, 10.432088},
           radius = 0.400009}
       }

  object = primobj{
         material = plastic { kd = .5, ks = 0.3, kt = 0,
         d_col = {0.6, 0.6, 1}, s_col = {1,1,1}},
         shape = sphere{
           center = {-0.649865, 0.321913, 10.492903},
           radius = 0.400009}
```

```
        }

  object = primobj{
        material = plastic { kd = .5, ks = 0.0, kt = 0,
        d_col = {0.6, 0.8, 0.6}, s_col = {1,1,1}},
        shape = sphere{
          center = {-0.593839, 1.181818, 9.336469},
          radius = 0.400009}
      }
};
```

# Bibliography

[1] Ron Brinkmann. *The Art and Science of Digital Compositing, Second Edition: Techniques for Visual Effects, Animation and Motion Graphics (Morgan Kaufmann Series in Computer Graphics).* Morgan Kaufmann Publishers Inc., San Francisco, CA, USA, 2 edition, 2008.

[2] Paul E. Debevec and Jitendra Malik. Recovering high dynamic range radiance maps from photographs. In *SIGGRAPH '97*, 1997.

[3] Paul E. Debeveck. Rendering synthetic objects into real scenes: Bringing traditional and image-based graphics with global illumination and high dynamic range photography. In *SIGGRAPH '98*, 1998.

[4] Frederic Devernay, Olivier Faugeras, and Inria Sophia Antipolis. Automatic calibration and removal of distortion from scenes of structured environments. In *SPIE, volume 2567*, 1995.

[5] Tim Dobbert. *Matchmoving: The Invisible Art of Camera Tracking.* Sybex, 2005.

[6] M. Galassi et al. *GSL Reference Manual—Third Edition.* 2009.

[7] Gerald E. Farin and Dianne Hansford. *The Geometry Toolbox for Graphics and Modeling.* A. K. Peters, Ltd., USA, 1998.

[8] Martin A. Fischler and Robert C. Bolles. Random sample consensus: a paradigm for model fitting with applications to image analysis and automated cartography. *Commun. ACM*, 24:381–395, 1981.

[9] David Forsyth and J. Ponce. *Computer Vision: A Modern Approach.* Prentice Hall, 01 2003.

[10] Helmut Fritzsche. *Programação Não-Linear.* Edgard Blucherd, São Paulo, Brazil, 1978.

[11] Simon Gibson, Jon Cook, Toby Howard, and Roger Hubbold. Accurate camera calibration for off-line, video-based augmented reality. In *ISMAR*, pages 37–46, 2002.

[12] Jonas Gomes, Luiz Velho, and Mario Costa. *Design and Implementation of 3D Graphics Systems.* Taylor and Francis, New York, 2012.

[13] Richard Hartley and Andrew Zisserman. *Multiple View Geometry in Computer Vision*. Cambridge University Press, New York, NY, USA, 2 edition, 2003.

[14] Richard I. Hartley. In defense of the eight-point algorithm. *IEEE Transactions on Pattern Analysis and Machine Intelligence*, pages 580–593, 1997.

[15] Berthold K. P. Horn. Tsai's camera calibration revisited. *MIT Thechnical Report*, 2000.

[16] Barry R. James. *Probabilidade: um curso em nivel intermediario (segunda edicao)*. IMPA, 1996.

[17] Henrique Wann Jensen. *Realistic Image Synthesis Using Photon Mapping*. 2001.

[18] Hugh Christopher Longuet-Higgins. A computer algorithm for reconstructing a scene from two projections. *Nature*, 293:133–135, 1981.

[19] Bruce D. Lucas and Takeo Kanade. An iterative image registration technique with an application to stereo vision. In *IJCAI81*, pages 674–679, 1981.

[20] Robert Osada, Thomas Funkhouser, Bernard Chazelle, and David Dobkin. Shape distributions. *ACM Transactions on Graphics*, 21(4):807–832, October 2002.

[21] Philippe Bekart, Philip Dutre, and Kavita Bala. *Advanced Global Illumination*. 2003.

[22] Ademir Ribeiro and Elizabeth Karas. *Otimizacao Continua: Aspactos Teoricos e Computacionais*. 2014.

[23] Chaman Sabharwal. Stereoscopic projections and 3d scene reconstruction. In *ACM/SIGAPP*, pages 1248–1257, 1992.

[24] Mike Seymour. Art of tracking part 1: History of tracking, 2004.

[25] Jianbo Shi and Carlo Tomasi. Good features to track, 1994.

[26] Peter Shirley and Steve Marschner. *Fundamentals of Computer Graphics*. A. K. Peters, Ltd., USA, 3rd edition, 2009.

[27] Peter Shirley and R. Keith Morley. *Realistic ray tracing - Second Edition*. A K Peters, 2003.

[28] Carlo Tomasi and Takeo Kanade. Detection and tracking of point features. Technical report, International Journal of Computer Vision, 1991.

[29] R. Y. Tsai and T. S. Huang. Uniqueness and estimation of three-dimensional motion parameters of rigid objects with curved surfaces. pages 112–118, January 1982.

[30] Roger Y. Tsai. An efficient and accurate camera calibration technique for 3d machine vision. In *Proceedings of IEEE Conference on Computer Vision and Pattern Recognition*, pages 364–374, 1986.

[31] Allan Watt. *3D Computer Graphics (2nd Edition).* Addison-Wesley Pub, 1993.

# Index

Note: *Italic* page numbers refer to figures.